The Molecular Pharmacology of Essential Oils

Targets, Pathways, and Mechanisms

Dr. Scott A. Johnson

The Molecular Pharmacology of Essential Oils: Targets, Pathways, and Mechanisms / Scott A. Johnson

Cover design: Scott A. Johnson
Cover Copyright © 2025, by Scott A. Johnson

ISBN-13: 979-8988720669

Published by Scott A. Johnson Professional Writing Services, LLC: Orem, UT

Discover more books by Scott A. Johnson at authorscott.com/shop/

Contents

Essential Oil Polypharmacology—A Leap Forward in Therapeutics

Essential oils—volatile aromatic compounds extracted from plants—have been used for centuries in diverse cultures around the world for their numerous therapeutic properties and to maintain or restore mental, emotional, and physical well-being. These natural substances are complex mixtures of a dozen to hundreds of bioactive molecules that can exert multiple pharmacological actions—a characteristic known as polypharmacology. Based on our growing understanding of the diverse bioactivity of each essential oil, we can confidently state that essential oils are multi-target, multi-mechanism natural solutions with diverse therapeutic purposes and mechanisms. Recent advancements in phytochemistry, molecular biology and modeling, biochemistry, genomics and proteomics, and pharmacology have unveiled the multifaceted nature of essential oils and provided insights into their mechanisms of action, molecular targets, and biological pathways. By integrating this knowledge, we can enhance the effective application of essential oils as natural therapeutics, allowing for safer and more efficient use in clinical practice and holistic healthcare.

While clinical research ideally informs the use of essential oils—providing crucial insights into effective methods and dosages—sometimes such research is limited for specific conditions. In these cases, understanding the properties of essential oils, along with their molecular targets, can guide their application.

Molecular pharmacology is a branch of pharmacology that focuses on the study of the interactions between therapeutics and biological molecules at the molecular level. It examines how therapeutic agents influence biological processes by interacting with specific molecular targets such as proteins (e.g., receptors and enzymes), nucleic acids (DNA and RNA), and other cellular components. Overall, molecular pharmacology bridges the gap between classical pharmacology, which often focuses on the effects of therapeutics at the organism level, and molecular biology, providing a detailed understanding of how and why therapeutics work at the molecular and cellular levels. The application of molecular pharmacology to essential oils is a substantial leap forward in their therapeutic use.

For example, BRCA1 (Breast Cancer Gene 1) is a crucial tumor suppressor gene responsible for repairing damaged DNA and preventing uncontrolled cell growth. It plays a vital role in either repairing genetic damage that could lead to cancer or triggering cell death when repair is not possible. When BRCA1 is mutated, its ability to fix DNA damage is compromised, which increases the risk of developing certain cancers, including breast and ovarian cancer, as well as, to a lesser extent, pancreatic, prostate, and colon cancer.

Notably, research has indicated that alpha-santalol, the primary active constituent in sandalwood (*Santalum album, S. austrocaledonicum, S. paniculatum, S. spicatum*) essential oil, can enhance the expression of BRCA1 in both estrogen receptor-positive and estrogen receptor-negative breast cancer cells.[1] This finding suggests that alpha-santalol—and, by extension, sandalwood essential oil—may offer promising natural options to potentially reduce breast cancer risk in women with BRCA1 mutations. This example illustrates how a deep understanding of the molecular targets associated with essential oils can inform their use in addressing various health concerns for both prevention and treatment.

Polypharmacology refers to the ability of a single agent or compound to interact with multiple targets through various mechanisms. This is particularly salient in the context of essential oils, as they are not solitary chemicals but rather complex mixtures that act in an additive, synergistic, and antagonistic manner. The therapeutic potential of these compounds lies in their ability to modulate multiple biological pathways simultaneously. While many traditional pharmaceutical drugs are designed to contain a single active compound, which is specifically developed and tested for a particular therapeutic effect, essential oils are complex mixtures of bioactive constituents each contributing to the overall therapeutic profile. This complexity can lead to synergistic effects, where the combined action of constituents is greater than the sum of their individual effects. It can also reduce the potential unwanted effects of a constituent in essential oils that is toxic in isolation. In other words, essential oils were divinely designed as multi-target, multi-mechanism agents that self-regulate any potentially harmful effects when used within reasonable guidelines. Now, to be clear, this does not mean that essential oils cannot cause an unwanted effect. This is contrary to the basic principles of Paracelsus's Law, which encapsulates the idea that the therapeutic potential of any substance involves a careful balance with the risk of toxicity, highlighting the importance of dosage and exposure in determining effects. Therefore, understanding the intricate balance between their therapeutic benefits and potential risks is crucial for harnessing the full power of essential oils in a safe and effective manner.

The essential oil of lavender (*Lavandula angustifolia*) is an excellent example of polypharmacology. It contains numerous compounds such as linalool, linalyl acetate, (Z)-beta-Ocimene, beta-caryophyllene, and lavandulyl acetate that exhibit various pharmacological activities, including anxiolytic, sedative, and anti-inflammatory effects. Indeed, lavender oil typically contains between one hundred and two hundred different constituents, frequently with around fifty constituents, making up over ninety-five percent of its total composition. Evidence suggests that these constituents can influence neurotransmitter levels, demonstrating a multifaceted approach to alleviating anxiety and promoting relaxation.[2,3,4] Unlike synthesized drug pharmacotherapy, which often targets a single pathway and may lead to undesired side effects because it lacks buffering antagonistic compounds, essential oils can provide a more balanced therapeutic profile through their polypharmacological nature.

Understanding the mechanisms by which essential oils exert their effects is crucial for leveraging their therapeutic potential. Key mechanisms include:

Receptor Modulation—Essential oil constituents can interact with various receptors in the body. For instance, linalool from lavender can bind to gamma-aminobutyric acid (GABA) receptors,[5] enhancing inhibitory neurotransmission. This interaction can lead to reduced anxiety and stress. Similarly, other compounds in essential oils—benzyl benzoate, benzyl alcohol, alpha-santalol, borneol—may affect serotonin receptors,[6,7,8] potentially influencing mood and emotional well-being. Benzyl benzoate in ylang ylang (*Cananga odorata*) activates both serotonin and dopamine pathways, which reduce anxiety.[9] Furthermore, bergamot (*Citrus bergamia*) essential oil modulates metabotropic glutamate receptors (mGluRs), which partly explains its impact on behavior. mGluRs are a specific type of glutamate receptors that play a crucial role in regulating synaptic transmission, neuronal excitability, and plasticity in the central nervous system. Distinct from ionotropic glutamate receptors (AMPA, NMDA, and kainate receptors), which directly mediate fast synaptic transmission, mGluRs are G-protein-coupled receptors (GPCRs) that work through intracellular signaling pathways to fine-tune neuronal activity. These receptors are targets for neurodegenerative diseases, mental health disorders, chronic pain, and addiction. Again, because essential oils will target multiple receptors simultaneously due to their complexity, they will tend to be more balancing and have a lower side effect risk than drugs.

Notably, sesquiterpenes generally have greater binding potential and affinity than monoterpenes when interacting with human cell receptors. This is due to several key factors related to their chemical structure and molecular interactions.

Monoterpenes—

- Monoterpenes ($C_{10}H_{16}$, example: limonene) are smaller, more volatile, and have simpler structures, allowing them to easily cross membranes but sometimes leading to weaker or shorter interactions with receptors.

- Monoterpenes often lack oxygenated groups (e.g., ketones, lactones, alcohols), meaning they exert effects via transient interactions or metabolic conversion.
- Monoterpenes are usually more volatile and metabolized faster, leading to shorter durations of action.
- Monoterpenes can still exert significant biological effects, often through indirect mechanisms like enzyme modulation or second messenger signaling.

Sesquiterpenes—

- Sesquiterpenes ($C_{15}H_{24}$, example: beta-caryophyllene) are larger and more complex, often featuring additional functional groups (e.g., hydroxyl, ketone, lactone), which can enhance their ability to bind with greater affinity to receptor sites.
- Many sesquiterpenes contain oxygenated groups, which allow for stronger hydrogen bonding and van der Waals interactions with receptors.
- Sesquiterpenes tend to have longer half-lives in the body, meaning their effects last longer.
- Due to their larger size, lipophilicity, diverse functional groups, and stronger receptor interactions, sesquiterpenes generally have higher binding affinity than monoterpenes when interacting with human cell receptors.

Inflammatory Pathways—Essential oils frequently demonstrate anti-inflammatory properties. For example, copaiba (*Copaifera* spp.) essential oil can inhibit the production of proinflammatory cytokines, such as tumor necrosis factor-alpha (TNF-α) and 5-lipoxygenase (5-LOX).[10,11] This action can be beneficial for conditions characterized by chronic inflammation, such as arthritis and inflammatory bowel diseases. Additionally, oils such as tea tree (*Melaleuca alternifolia*) and ginger (*Zingiber officinalis*) essential oil have been found to modulate inflammatory responses, further supporting their use in various inflammatory conditions.[12,13] In fact, for all the major human inflammatory targets, TNF-α, 5-LOX, interleukin-1 beta (IL1-β), interleukin 6 (IL-6), interleukin 8 (IL-8), interferon-gamma (IFN-γ), inducible nitric oxide synthase (iNOS), cyclooxygenases (COX-1 and COX-2), nuclear factor-κB (NF-κB), inflammasomes, prostaglandins, leukotrienes, reactive oxygen species (ROS), and the complement system, there is an essential oil that attacks every one of these pathways.[14]

Inflammation is regulated by several key molecular targets in the human body, including enzymes, cytokines, receptors, and signaling pathways. Here are the most important ones:

Proinflammatory Cytokines & Chemokines (TNF-α, IL1-β, IL-6, IL-8, IFN-γ)— signaling molecules that mediate inflammation by activating immune cells.

- TNF-α: Master regulator of inflammation, promoting immune cell activation and tissue damage; a key cytokine involved in various autoimmune and autoinflammatory conditions.

[10]

- IL1-β: Triggers fever and amplifies inflammation.
- IL-6: Drives acute and chronic inflammation; key in autoimmune diseases.
- IL-8: A chemokine that recruits neutrophils to inflamed areas.
- IFN-γ: Stimulates immune responses, particularly against infections; dysregulation or chronic elevation can lead to excessive inflammation and contribute to autoimmune and inflammatory diseases.

Inflammatory Enzymes (COX-1, COX-2, LOX, iNOS)—These enzymes drive the production of inflammatory molecules.

- COX-1 and COX-2: Convert arachidonic acid into prostaglandins, which cause pain, fever, and swelling.
- LOX: Produces leukotrienes, which contribute to asthma and allergic inflammation. Aberrant 5-LOX activity and the resulting leukotriene production have been implicated in various autoimmune diseases and inflammatory conditions.
- iNOS: Generates excess nitric oxide (NO), leading to oxidative stress and inflammation.

Nuclear Factor-κB (NF-κB) Pathway—A is a key transcription factor that activates genes for proinflammatory cytokines, chemokines, and enzymes. Chronic NF-κB activation is linked to autoimmune diseases, cancer, and chronic inflammation.

Inflammasomes (NLRP3)—The NLRP3 inflammasome detects cellular stress (infections, toxins, damage) and activates IL-1β and IL-18, leading to inflammation. Overactivation contributes to autoimmune diseases, gout, neuroinflammation, and metabolic disorders.

Prostaglandins & Leukotrienes—

- Prostaglandins (PGE_2, PGD_2, $PGF_2\alpha$, PGI_2): Mediate pain, fever, and swelling.
- Leukotrienes (LTB_4, LTC_4, LTD_4): Drive allergic inflammation, bronchoconstriction, and asthma.

Reactive Oxygen Species (ROS) & Oxidative Stress—ROS (like superoxide and hydrogen peroxide) damage cells, activate inflammatory pathways, and drive chronic diseases. Excess ROS contribute to neurodegeneration (Alzheimer's, Parkinson's), cardiovascular disease, and arthritis.

Complement System—A group of proteins (C3, C5, etc.) that amplify inflammation, recruit immune cells, and destroy pathogens. Overactivation is linked to autoimmune diseases, sepsis, and inflammatory damage.

Antioxidant Activity—Essential oils like oregano (*Origanum vulgare*), clove (*Eugenia caryophyllata*, syn. *Syzigium aromaticum*), and thyme (*Thymus vulgare*, *Th. zygis*) possess potent antioxidant properties due to their high content of phenolic compounds. By scavenging free radicals and mitigating oxidative stress, these oils can help protect cells from damage and support overall health. Other oils, such as cinnamon bark (*Cinnamomum verum*) and rosemary (*Rosmarinus officinalis*), also exhibit significant antioxidant capabilities, which makes them valuable for preventing oxidative damage related to chronic diseases.

Essential oils that contain polyphenolic compounds—clove, oregano, thyme, and cinnamon—often exhibit high antioxidant capacity due to the particular chemical structure and properties of these compounds. Polyphenols have multiple phenolic hydroxyl groups (-OH) in their chemical structure. These hydroxyl groups are crucial for their antioxidant effects because they can donate hydrogen atoms or electrons to free radicals, neutralizing them and preventing oxidative damage to cells and tissues. Polyphenols have favorable redox (reduction-oxidation) properties, allowing them to undergo oxidation and reduction reactions. This capability enables them to act as reducing agents, helping to prevent the oxidation of other biomolecules, such as lipids, proteins, and DNA. Some polyphenols can influence the activity of antioxidant enzymes in the body, such as superoxide dismutase (SOD), catalase, and glutathione peroxidase. By enhancing the activity of these enzymes, polyphenols contribute to a more robust antioxidant defense system. These characteristics make polyphenol-rich essential oils valuable tools in both health promotion and disease prevention.

Lastly, many essential oils can increase the body's natural (endogenous) antioxidant defenses and activity. Essential oil compounds, like phenols and terpenes, can stimulate the body's own antioxidant defense mechanisms by promoting the production and activity of key enzymes such as SOD, catalase, and glutathione peroxidase. By increasing the capacity for endogenous antioxidant production, essential oils help neutralize harmful free radicals and reduce oxidative stress, which is linked to chronic diseases and aging. The benefit of this enhancement lies in the potential for improved cellular health, reduced inflammation, and a lower risk of conditions such as cardiovascular diseases, neurodegenerative disorders, and cancer.

Antioxidant activity is assessed via various assays:

DPPH (2,2-Diphenyl-1-picrylhydrazyl)—measures the ability of an antioxidant to donate electrons or hydrogen atoms to the DPPH free radical. It is a quick and simple method, useful for screening antioxidant activity but may not reflect the complex interactions occurring in biological systems.

FRAP (Ferric Reducing Antioxidant Power)—evaluates the ability of antioxidants to reduce ferric ions (Fe^{3+}) to ferrous ions (Fe^{2+}). This assay is specific for measuring the reducing power of antioxidants, primarily looking at their ability to reduce metal ions.

ABTS (2,2'-Azino-bis(3-ethylbenzothiazoline-6-sulfonic acid))—assesses the ability of antioxidants to scavenge the ABTS radical cation, which is produced by the oxidation of ABTS. The ABTS assay is applicable to both lipophilic and hydrophilic antioxidants, making it more versatile in various sample types.

TEAC (Trolox Equivalent Antioxidant Capacity)—measures the ability of antioxidants to scavenge the ABTS radical cation, similar to the ABTS assay. This method quantifies the antioxidant activity in terms of Trolox equivalents, making it easier to compare results across different samples.

Hydroxyl Radical Scavenging—measures the ability of a sample to scavenge hydroxyl radicals, one of the most reactive and damaging free radicals. It focuses specifically on hydroxyl radicals, providing insight into the capacity to neutralize highly reactive species.

Total Phenolic Content (TPC)—while not a direct measure of antioxidant capacity, TPC assays indicate the total amount of phenolic compounds in a sample, which are known for their antioxidant properties. TPC gives an estimate of the total phenolic content, while other assays provide specific measures of antioxidant activity.

ORAC (Oxygen Radical Absorbance Capacity)—measures the total antioxidant capacity of a sample by evaluating its ability to quench peroxyl radicals. It is designed to assess the protective capacity of antioxidants against oxidative stress in biological systems.

OXYGEN RADICAL ABSORBANCE CAPACITY (ORAC) SCORE OF CERTAIN ESSENTIAL OILS.[15]

Essential Oil	ORAC Score
Clove	1,078,700
Myrrh	379,800
Fennel	238,400
Thyme	157,380
Oregano	153,007

Antimicrobial Action—The inherent antimicrobial properties of many essential oils are attributable to their capacity to disrupt microbial cell membranes, inhibit biofilm formation, and compromise metabolic processes in pathogens. For instance, tea tree oil (*Melaleuca alternifolia*) exhibits significant activity against various bacteria, fungi, and viruses, making it a valuable agent in managing infections.[16] Moreover, the essential oil from thyme has been shown to be effective against antibiotic-resistant strains of bacteria, highlighting its role in modern antimicrobial strategies.[17] Some essential oils—geranium (*Pelargoneum graveolens*), oregano, thyme, lemongrass, rosemary—even increase the effectiveness of antibiotics, making them possible complementary solutions that could reduce the dose of antibiotic required.[18]

Importantly, pathogenic microbes generally do not build up a resistance to essential oils in the same manner as they do to conventional antibiotics, primarily due to the inherent variability in composition of essential oils from batch to batch. Each essential oil is a complex mixture of numerous volatile compounds, which can vary significantly based on factors such as the plant source, extraction method, and environmental conditions. This variability means that microbes are continually faced with different chemical profiles, making it challenging for them to adapt or develop resistance mechanisms specific to a single component. Additionally, essential oils can exhibit a multicomponent action mechanism, where multiple phyto-chemicals may simultaneously target different sites within microbial cells, disrupting various vital processes such as cell membrane integrity, metabolic pathways, and genetic material synthesis. This multifaceted mode of action not only reduces the likelihood of resistance development but can also result in a synergistic effect, enhancing the antimicrobial efficacy of the essential oils. Consequently, the dynamic and complex nature of essential oils serves as a significant barrier to microbial resistance, highlighting their potential as alternative antimicrobial agents.

Microbiome Modulation—Essential oils can also influence the human microbiome, promoting the growth of beneficial microorganisms while inhibiting pathogenic ones. Based on emerging research surrounding essential oils, the definition of a prebiotic needs to be broadened beyond merely non-digestible fibers that probiotics ferment. While the traditional definition emphasizes specific types of dietary fibers (such as inulin and oligofructose) that selectively stimulate the growth and activity of beneficial gut bacteria, the concept can encompass a wider range of substances that create a favorable environment for probiotics in the human gut. This broader definition includes compounds that enhance the overall gut environment, contribute to maintaining a healthy gut barrier, or trigger postbiotic production (e.g., butyrate). This comprehensive perspective recognizes the diversity of mechanisms that support probiotic activity and gut health, emphasizing the complex interplay between diet, microbiota, and host physiology.

Some essential oils appear to act as a prebiotic and trigger the production of postbiotics, by creating a favorable gut environment for healthy microbes to thrive. This is the case even when the oils are used at concentrations that still permit them to inhibit harmful microbes. Some essential oils demonstrating this effect include lemongrass (*Cymbopogon flexuosus*), peppermint (*Mentha piperita*), and tangerine (*Citrus reticulata* Blanco). These findings are both fascinating and exciting and have massive implications for using essential oils to support gut health and the gut microbiome, and producing far-reaching benefits for inflammatory balance, immune function, cognitive function, and even mood regulation.

Enzyme Interactions—Some essential oils can influence the activity of various enzymes in the body, participating in metabolic processes or modulating detoxification pathways. They possess a diverse array of bioactive compounds that can significantly influence human enzyme activity, impacting various physiological processes and health outcomes. For instance, essential oils such as rosemary, cinnamon, and peppermint have been shown to inhibit enzymes like acetylcholinesterase (AChE) and butyrylcholinesterase (BChE),[19, 20, 21] which play critical roles in neurotransmission by breaking down acetylcholine in the synaptic cleft. This inhibition could enhance cognitive function and memory, suggesting potential benefits for neurodegenerative conditions like Alzheimer's disease. Similarly, essential oils like clary sage (*Salvia sclarea*) and geranium can inhibit tyrosinase, an enzyme involved in melanin production, thereby potentially aiding in skin lightening and the management of hyperpigmentation.[22,23] Furthermore, essential oils such as helichrysum (*Helichrysum italicum*) exhibit inhibitory effects on elastase, an enzyme that contributes to skin aging through the degradation of elastin.[24] This can promote skin health and elasticity.

On the metabolic front, essential oils like those from lime (*Citrus limon*), lemon (*Citrus aurantifolia*), and grapefruit (*Citrus paradisi*) have been shown to inhibit alpha-amylase and alpha-glucosidase, enzymes involved in carbohydrate digestion, which may help regulate blood sugar levels and support metabolic health.[25,26] Additionally, research shows allspice (*Pimenta dioica*) essential oil can modulate lipase activity, influencing fat digestion and absorption, thus contributing to weight management and cardiovascular health.[27] Overall, the ability of essential oils to interact with these enzymes highlights their potential therapeutic benefits, ranging from cognitive enhancement to metabolic regulation and skin health, underscoring their value in holistic health approaches.

Essential oils interact with human enzymes through several mechanisms, primarily by binding to active sites, altering enzyme activity, and modifying protein structures. These interactions depend on the chemical composition of the essential oil, enzyme specificity, and molecular docking compatibility. Many essential oil compounds act as competitive inhibitors, meaning they mimic the enzyme's natural substrate and bind to its active site, preventing normal enzyme function. For instance, 1,8-cineole competes with acetylcholine for the active site, slowing its breakdown and enhancing neurotransmission.[28] Some essential oil compounds bind to a different site (allosteric site) on the enzyme, changing its shape and reducing activity without directly competing with the substrate. Thymol is a positive allosteric modulator of human GABA(A) receptors, enhancing the receptor's response to its natural ligand, gamma-aminobutyric acid (GABA), without directly mimicking its action at low concentrations.[29] Certain essential oils influence enzymes by affecting oxidative stress levels, which can either inhibit or enhance enzyme function. Essential oils with strong antioxidant properties may reduce elastase activity, slowing wrinkle formation and skin aging.[30,31] Some enzymes, like tyrosinase and amylase, require metal ions (e.g., copper, zinc) to function. Essential oils with chelating properties remove these metals, leading to enzyme inhibition. Allspice chelates metals and therefore could inhibit both amylase and tyrosinase, improving metabolism and the appearance of the skin.[32] Lastly, some essential oil constituents interact with hydrogen bonds and van der Waals forces that stabilize enzyme-substrate interactions, effectively disrupting normal function. Put simply, some constituents in essential oils can disrupt the weak bonds that allow enzymes to connect with and act on other molecules. This interference can prevent the enzyme from performing its function. For example, eugenol can weaken the connection between the amylase enzyme and starch, slowing down the release of glucose from starch, thus helping regulate glucose spikes after eating.[33] Overall, essential oils influence human enzymes through diverse mechanisms, highlighting their potential therapeutic applications in areas such as neuroprotection, skin health, metabolic regulation, and anti-aging.

Modulation of Nervous System Activity—Essential oils (EOs) have the ability to influence the nervous system not just by modulating receptors, but also through complex biochemical pathways that can regulate neurotransmitter levels, brain activity, and emotional responses.

Essential oils can affect neurotransmitters—the chemicals responsible for transmitting signals in the brain and nervous system. By interacting with the receptors that control neurotransmitter release, they can either increase or decrease neurotransmitter levels, leading to calming or stimulating effects. German chamomile (*Matricaria chamomilla*) contains the sesquiterpene alpha-bisabolol, which binds to GABA receptors (the brain's "calming" receptors) much like benzodiazepines (anti-anxiety medications). This promotes relaxation by reducing neuronal firing and inhibiting overactive brain areas, leading to reduced feelings of stress and anxiety.[34] Bergamot oil has antidepressant-like effects due to its ability to modulate the serotonergic and dopaminergic systems. Specifically, linalool in bergamot has been shown to increase serotonin release, which can help elevate mood and reduce anxiety.[35,36] The interaction of essential oils with neurotransmitter systems highlights their potential as natural therapeutic agents for promoting relaxation, reducing anxiety, and enhancing overall emotional well-being, particularly considering their ability to cross the blood-brain barrier.

Essential oils can influence brainwave patterns, enhancing focus, relaxation, or even sleep, depending on the oil. Certain compounds found in essential oils directly impact neural activity in the brain, increasing or decreasing specific brainwave frequencies. Menthol has been found to activate ion channels in the brain, particularly the TRPM8 receptors, which respond to cold and cooling sensations.[37,38] These receptors are involved in brain alertness and cognitive function. The stimulating effects of menthol in peppermint may increase beta waves, associated with alertness, concentration, and cognitive performance. Interestingly, ravintsara (*Cinnamomum camphora*), an essential oil rich in 1,8-cineole and typically known for its stimulatory effects on the brain, notably reduces absolute beta, absolute gamma, absolute high beta, and relative gamma brainwaves, while simultaneously increasing the relative theta and power spectra values. These observations suggest inhaling ravintsara oil promotes mental and physical relaxation, enhances cognitive processes related to memory and attention, and enhances mood states. In conclusion, the diverse effects of essential oils on brainwave patterns underscores their potential as natural tools for enhancing mental alertness, relaxation, and cognitive function, offering a sensory approach to promoting overall well-being.

The autonomic nervous system (ANS) controls involuntary functions like heart rate, digestion, and respiratory rate. Essential oils can influence the sympathetic (fight-or-flight) and parasympathetic (rest-and-digest) systems, either calming the body or promoting a more alert, energetic state. Inhalation of lemon oil balances the ANS. It differentially influenced ANS activity in healthy and depressed persons—sympathetic nervous system activity is often elevated and parasympathetic activity decreased in people who are depressed. Healthy subjects experienced increased sympathetic and parasympathetic activity after inhalation, whereas depressed persons experienced increased parasympathetic activity.[39] The ability of essential oils to modulate autonomic nervous system activity highlights their therapeutic potential for promoting balance and resilience in both healthy individuals and those experiencing depression or other conditions involving dysregulated ANS function.

Chronic neuroinflammation is implicated in many neurodegenerative diseases and emotional disorders. Essential oils with anti-inflammatory properties can help reduce inflammation in the brain, improving mood, cognition, and overall nervous system health. One promising essential oil for neuroinflammation is turmeric (*Curcuma longa*) because of its high *ar*-turmerone content. *Ar*-Turmerone is a volatile sesquiterpene ketone and considered the major bioactive constituent of turmeric essential oil. Promising preclinical research demonstrated that ingestion of *ar*-turmerone provided protection against brain damage caused by neuroinflammation. The constituent normalized brain glucose utilization, inhibited microglial cell activation (nerve cells that act as key mediators of neuroinflammation and are important to triggering the brain degeneration process), and suppressed the production of inflammatory cytokines.[40] Other research shows that *ar*-turmerone significantly increases the production of neural stem cells—brain stem cells that are essential for the repair and recovery of brain function.[41] Other essential oils that protect against damage caused by neuroinflammation include copaiba, ginger, and thyme.[42, 43, 44] The multi-mechanisms—antioxidant, modulation of inflammatory mediators, influence on microglia, regulation of neurotransmitters, activation of anti-inflammatory pathways (Nfr2), reduction of stress and anxiety, and direct interaction with neurons—by which essential oils attack neuroinflammation and their ability to rapidly enter brain tissues, make them leading candidates for conditions involving neuroinflammation.

As a side note, during sleep, the brain undergoes crucial restorative processes. One of the most important of these is the clearance of metabolic waste products, which are typically accumulated during waking hours. This process is partly facilitated by the glymphatic system—a network that helps remove waste and toxins from the brain, primarily during sleep. This process requires glucose to power the brain's activity. If you are someone who wakes up regularly between the hours of 3 and 5 a.m., your brain may be short on glucose to perform this process. When the brain lacks sufficient glucose during sleep, it triggers a stress response, leading to the release of cortisol and adrenaline to mobilize stored glucose. This stress response can also activate histamine release, particularly in individuals prone to histamine intolerance or mast cell activation. Adrenaline and cortisol can stimulate mast cells, leading to increased histamine levels. Unfortunately, these hormones also wake you up. To prevent this, have a small snack right before bed of something like raw honey in MCT oil, or honey with collagen or amino acids.

Another pathway by which essential oils influence nervous system activity is the olfactory pathway. The olfactory system plays a crucial role in how essential oils affect nervous system activity due to its direct connections with various brain regions involved in emotion, memory, and autonomic functions. When essential oils are inhaled, volatile compounds interact with olfactory receptors in the nasal epithelium—the thin layer of cells that lines the inside of your nose—which then send signals to the olfactory bulb. This structure processes the olfactory information and relays it to several brain regions, including the limbic system (which encompasses the amygdala and hippocampus), hypothalamus, and even the prefrontal cortex. The limbic system is particularly significant because it is involved in emotional regulation, memory formation, and the body's response to stress, making it

a critical target for the rapid mood-modulating effects of essential oils.

Once the olfactory receptors detect the specific compounds in essential oils, they activate G protein-coupled receptors (GPCRs), leading to a series of intracellular signaling events. This process can also influence neural pathways related to autonomic functions. For instance, the activation of olfactory receptors can initiate a cascade of signaling that affects neurotransmitter release, such as serotonin, dopamine, and norepinephrine, which play critical roles in mood regulation. By modulating these neurotransmitter levels, essential oils can create calming or uplifting effects, depending on the specific oil and its molecular composition.

The olfactory system is not a singular entity but rather a collection of distinct components categorized by their anatomy, function, and evolutionary history. The primary division is the pure olfactory system, also known as the main olfactory system, which is responsible for detecting general odorants in the environment. Essential oils primarily interact with the pure olfactory system. This system relies on the olfactory epithelium in the nasal cavity, where olfactory receptor neurons (ORNs) detect odor molecules and transmit signals to the olfactory bulb for initial processing. From there, the processed signals travel via the olfactory tract to the olfactory cortex, which includes areas like the piriform cortex, amygdala, and entorhinal cortex. This pathway allows for the conscious perception of a wide array of airborne chemicals and establishes strong links between smell, memory, and emotion through its connections with the limbic system.

In contrast, the accessory olfactory system, or vomeronasal system, is a more specialized system primarily involved in the subconscious detection of social and reproductive chemical cues, such as pheromones. Its key component is the vomeronasal organ (VNO), which detects these cues and sends signals to the accessory olfactory bulb, with connections primarily to limbic areas like the hypothalamus and amygdala. While its role is significant in animal behavior, its function in adult humans remains a topic of debate.

Finally, the trigeminal chemosensory system, though not strictly part of the olfactory system, often works in conjunction with it. This system is stimulated by irritating or strong-smelling substances and is responsible for detecting sensations like pungency, burning, or cooling, contributing to a more complete sensory experience of certain substances.

In addition to neurotransmitter modulation, essential oils can target various molecular pathways involved in the body's stress response. Some compounds found in essential oils, such as linalool and limonene, have been shown to downregulate hyperactive hypothalamic-pituitary-adrenal (HPA) axis activity. The HPA axis regulates the release of cortisol, the primary stress hormone, and dysregulation of this axis is associated with anxiety and depression. By attenuating this stress response, essential oils can help reduce neuroinflammation and promote a more balanced emotional state.

Further molecular interactions can occur at the level of inflammation in the nervous system. Essential oils can inhibit the activity of proinflammatory cytokines—proteins released by immune cells that can exacerbate neuroinflammation. For instance, compounds like eugenol from clove oil or beta-caryophyllene from copaiba have demonstrated the ability to modulate inflammatory mediators, acting on pathways such as nuclear factor kappa B (NF-κB). By inhibiting NF-κB activation, essential oils can reduce the expression of inflammatory cytokines, leading to a decrease in neuroinflammation and potentially providing neuroprotective effects.

Ultimately, the multifaceted interactions between the olfactory system and essential oils illustrate their potential for influencing nervous system activity. Through modulation of neurotransmitter systems, regulation of the stress response, and inhibition of neuroinflammatory pathways, essential oils can effectively alter mood and emotional states, showcasing their therapeutic promise in mental wellness, cognitive health, and other processes controlled by the nervous system. Further research is needed to completely map these interactions and identify specific molecular targets, but the existing evidence underscores the significant role of olfactory stimuli in regulating complex neural processes.

Immune Modulation—The immune system is a complex network of cells and organs that protect the body. Measuring its function is also a complex task, as it involves evaluating various aspects of the immune system, including immune cell numbers, activity, and the balance of proinflammatory and anti-inflammatory responses. Here are some key markers to assess immune function:

Complete Blood Count (CBC) with Differential—A complete blood count provides a basic overview of immune cell populations, including total white blood cells and their subtypes (lymphocytes and neutrophils). Elevated lymphocyte counts may indicate viral infections or chronic inflammation. Insufficient lymphocytes mean you have an increased risk of infection and impaired immune response. Increased neutrophil counts often indicate bacterial infections or acute inflammation. A low neutrophil count increases the risk of bacterial infections and delays responses to injury and infection.

Lymphocyte Subset Analysis—Flow cytometry is often used to identify and quantify specific lymphocyte populations. T cells (CD4+ and CD8+) are central to adaptive immunity, with CD4+ T helper cells supporting B cells and regulating other immune cells, while CD8+ T cells directly kill infected or cancerous cells. The balance between CD4+ and CD8+ T cells is crucial for maintaining a properly functioning immune system. These two types of T lymphocytes play different but complementary roles in immune defense, and an imbalance between them can lead to immunodeficiencies, autoimmune diseases, or chronic infections. B cells are responsible for antibody production. Natural killer cells (NKCs) are involved in the innate immune response, particularly against tumor cells and virally infected cells. Analyzing these subsets provides insights into immune function and balance.

Cytokine Profiling—Cytokines are signaling molecules that mediate and regulate immunity and inflammation. Assessing the levels of key cytokines can provide valuable information about immune function, particularly your inflammatory response.

Immunoglobulin Levels—Measuring the levels of different classes of antibodies (immunoglobulins, such as IgG, IgA, and IgM) provides insights into humoral immunity. IgG is the most abundant antibody and important for long-term immunity. Playing a critical role in mucosal immunity (e.g., gut, genitourinary tract, and respiratory tract), IgA helps defend against infections in vulnerable areas where pathogens are most likely to enter. IgM is typically the first antibody produced in response to infection and a powerful defender during the early stages of the immune response.

Inflammation Markers—Markers such as C-reactive protein (CRP), erythrocyte sedimentation rate (ESR), and fibrinogen can indicate systemic inflammation and immune activity. Elevated levels can suggest chronic inflammatory conditions or immune dysregulation.

Autoantibodies—In cases where autoimmune disorders are a concern, measuring specific autoantibodies (e.g., anti-nuclear antibodies, rheumatoid factor) can indicate an inappropriate immune response, which may compromise immune balance.

Microbiome Analysis—Roughly seventy percent of your immune system is housed within your gut, called the gut-associated lymphoid tissue. Though not a direct measure of immune function, evaluating the gut microbiome's composition can offer insights into immune health. The microbiome plays a key role in educating the immune system, and its balance can influence systemic immune responses.

We often take too narrow of a view on how essential oils impact immune function. This then leads to searching for oils that kill or suppress the activity of specific pathogens. However, your immune system is an incredible system that tirelessly defends you against hazards every single day. It does a marvelous job most of the time. It's when our lifestyle is not in order or we are under attack from too many threats at once that the system gets overwhelmed and needs some extra support in the form of nutraceuticals or essential oils.

Essential oils can have immune-modulatory effects, either enhancing or suppressing immune responses. For instance, eucalyptus oil is known to stimulate immune function, making it potentially beneficial during respiratory infections, while others like lavender may promote a balanced immune response, reducing hyperactivity in allergic conditions. Clove oil is a powerful immune-supportive oil that increases white blood cell count, but simultaneously decreases proinflammatory cytokines.[45, 46, 47] It also restores cellular and humoral immune responses after suppression by drugs. For immune balance, frankincense is a go-to oil. Dendritic cells (DCs) are specialized antigen-presenting cells and sentinel cells (first responder cells of the immune system found in body tissues) involved in regulating the immune response. They act as messengers between the innate and adaptive immune systems, processing and presenting antigen materials to T cells. Immature (undifferentiated) DCs are maintained in the tissues and facilitate immune tolerance—a mechanism whereby the immune system learns to discriminate self-antigens from foreign substances and a key to avoiding a hyperimmune response to environmental substances. Frankincense essential oil (*Boswellia sacra*) triggered the differentiation of monocytes into immature DCs without stimulating immature DCs to mature.[48] This suggests that frankincense may be able to produce DCs that promote immune tolerance in conditions such as autoimmune disorders, allergies, hypersensitivity reactions, and after organ transplantation. Based on the current research, essential oils support healthy immune function and both stimulate and suppress immune function based on needs.

Analgesic and Antinociceptive Activity—Pain relief is a highly sought-after property, but unfortunately, many of the drugs used for pain relief have significant side effects. The two most common over-the-counter (OTC) drugs used are acetaminophen and ibuprofen. Acetaminophen overdose is the number one cause of liver failure in many developed countries. High doses can also lead to kidney dysfunction, while chronic high-dose use may increase cardiovascular complications (high blood pressure and heart disease). Ibuprofen is known to cause gastrointestinal issues, like ulcers and bleeding, especially after long-term use. It too can cause kidney damage and increase the risk of heart attack and stroke. Due to these adverse effects, people often turn to alternatives for pain relief.

Antinociception (blocks physical pathway of pain) is the reduction or elimination of pain at the level of the nervous system by blocking transmission of pain signals from the injury site to the brain. Analgesia (targets the brain's interpretation of pain) alters the brain's perception of pain signals to reduce pain.

Several essential oils possess analgesic properties that can help alleviate pain. For example, peppermint oil contains menthol, which acts as a natural analgesic and can provide relief from headaches and muscle pain. Similarly, oils such as lavender and rosemary have been used in aromatherapy for their pain-relieving effects and can be incorporated into topical formulations for localized relief. These effects are produced via multiple mechanisms or pathways relevant to essential oils:

Opioid pathway: These solutions exert their pain-relieving effects primarily through the activation of opioid receptors (mu, delta, and kappa) in the central nervous system. Activation of the mu-opioid receptor leads to inhibition of pain transmission pathways in the spinal cord and brain, decreasing the perception of pain.

Essential oils that work on this pathway: Atlas cedarwood (*Cedrus atlantica*), chaste tree (*Vitex agnus-castus*), eucalyptus (*E. globulus*), copaiba (beta-caryophyllene), lavandin (*Lavandula intermedia*), lavender, lemon, lemon eucalyptus (*Corymbia citriodora*), melissa (*Melissa officinalis*), pink pepper (*Schinus teribinthifolious*), and rosemary

Inflammatory pathways: Anti-inflammatories mainly work by inhibiting cyclooxygenase (COX) enzymes (COX-1 and COX-2), which play key roles in the conversion of arachidonic acid to prostaglandins. Prostaglandins are involved in inflammation and pain signaling. By limiting prostaglandin synthesis, anti-inflammatories reduce swelling and pain. This decreases the overall nociceptive input that reaches the central nervous system, helping to diminish both acute and chronic pain.

Essential oils that work on this pathway: Anise (*Pimpinella anisum*), caraway (*Carum carvi*), cassia (*Cinnamomum cassia*), cumin (*Cuminum cyminum*), copaiba (beta-caryophyllene, alpha-humulene), geranium, German chamomile, ginger, grapefruit, hinoki (*Chamaecyparis obtusa*), hyssop (*Hyssopus officinalis*) CT 1,8-cineole, lemongrass (*Cymbopogon citratus*), lemon, lemon eucalyptus, orange (*citrus sinensis*), marjoram (*Origanum majorana*), ocotea (*Ocotea quixos*), parsley (*Petroselinum crispum*), Roman chamomile (*Chamaemelum nobile*), rose (*Rosa damascena*), spearmint (*Mentha spicata*), tangerine (*Citrus nobilis*), Turkish oregano (*Origanum onites*), turmeric (*ar-*

Turmerone), and yarrow (*Achillea millefolium*)

Cannabinoid pathway: Cannabinoid solutions act on the endocannabinoid system by binding to cannabinoid receptors (CB1 and CB2). Activation of these receptors modulates pain perception, decreases inflammation, and influences mood and anxiety—an important aspect of pain management since our emotional responses influence our pain perception.

Essential oils that work on this pathway: Atlas cedarwood, black pepper (*Piper nigrum*; beta-caryophyllene), cananga (*Cananga odorata* f. *macrophylla*), copaiba (beta-caryophyllene), hemp (*Cannabis sativa*; selina-3,7(11)-diene), lavender, melissa (beta-caryophyllene), ylang ylang (beta-caryophyllene)

GABAergic pathways: These solutions enhance the effects of the neurotransmitter GABA, which inhibits neural activity. Increased GABAergic activity may dampen pain-signaling pathways, especially neuropathic pain.

Essential oils that work on this pathway: Anise, lemongrass, neroli (*Citrus aurantium; C. sinensis*), and peppermint

Serotonergic pathway: Some solutions increase levels of serotonin, a neurotransmitter involved in pain modulation. Serotonin can inhibit pain pathways in the spinal cord and brain, enhancing pain relief.

Essential oils that work on this pathway: Bergamot, frankincense (*B. sacra*)

Immune responses: Pain can lead to activation of the immune system, further exacerbating inflammation and pain sensations. Certain solutions can dampen this hyperactive immune activation and reduce inflammation to reduce pain.

Essential oils that work on this pathway: Many essential oils, but some of the top to consider are citruses, copaiba, frankincense, ginger, helichrysum, rosemary, and wintergreen.

While acetaminophen and ibuprofen are popular OTC pain relievers with known effective doses, their notable side effects often prompt individuals to seek alternative solutions. Incorporating essential oils into pain management can offer a natural and multifaceted approach to alleviating discomfort through various biochemical pathways. Oils like peppermint, lavender, and copaiba not only provide analgesic and anti-inflammatory benefits but also engage with the body's complex pain-signaling processes. As people increasingly prioritize holistic health solutions, exploring these natural options can empower them to manage pain effectively while mitigating the risks associated with traditional pharmaceuticals.

Respiratory Support: Respiratory health is vital to overall well-being, and its maintenance involves a complex interplay of various molecular targets, signaling pathways, and mechanisms. As respiratory diseases such as asthma, chronic obstructive pulmonary disease (COPD), and pulmonary fibrosis become increasingly prevalent, understanding the fundamental biological processes behind respiratory health is essential for developing effective treatments. Many essential oils, such as eucalyptus and peppermint, are known for their respiratory-supporting properties. These oils can help open airways, reduce inflammation in the respiratory tract, and promote easier breathing, making them valuable in managing conditions like asthma or allergies.

Inflammatory Pathways—Inflammation is a hallmark of many respiratory diseases. The regulation of inflammatory responses involves several significant molecular players:

Cytokine Signaling: Proinflammatory cytokines, such as interleukin-1 (IL-1), interleukin-6 (IL-6), and tumor necrosis factor-alpha (TNF-α), are pivotal in the pathogenesis of respiratory conditions. These cytokines mediate communication between immune cells and can escalate inflammation, leading to tissue damage. Targeting these cytokines or their receptors offers a therapeutic approach to managing inflammation and improving respiratory function.

Nuclear Factor kappa B (NF-κB): NF-κB is a transcription factor that regulates the expression of numerous genes responsible for the inflammatory response. Its activation is closely linked to the progression of respiratory diseases characterized by chronic inflammation. Inhibition of NF-κB signaling pathways can provide a strategic avenue for reducing inflammation in the lungs.

Essential oils that work on this pathway (specific to lung inflammation): bergamot, cinnamon bark, copaiba, eucalyptus (1,8-cineole), fennel (trans-anethole), frankincense, hinoki, and thyme

Oxidative Stress Pathways—Oxidative stress contributes significantly to respiratory dysfunction through the generation of reactive oxygen species (ROS).

- Reactive Oxygen Species (ROS): Excessive production of ROS can lead to cellular damage and inflammation in lung tissues. Antioxidants that neutralize ROS may help protect lung tissues from oxidative damage, thereby supporting respiratory health.
- Nrf2 Pathway: The Nrf2 pathway plays a protective role in the lungs by promoting the expression of antioxidant proteins and detoxifying enzymes. Activation of Nrf2 is a promising strategy to enhance the lung's defense against oxidative damage and inflammation.

Essential oils that work on this pathway: Catnip (*Nepeta cataria*), ginger, lavender, lemongrass (citral), patchouli (*Pogostemon cablin*; pogostone), rose, rosemary CT 1,8-cineole, spikenard, tea tree, thyme, yuzu (*Citrus junos*; limonene), and yarrow.

Mucin Production and Airway Remodeling—Mucin overproduction is associated with various respiratory disorders and can impede airflow, leading to symptoms such as coughing, wheezing, and shortness of breath. Conditions like asthma, chronic obstructive pulmonary disease (COPD), and cystic fibrosis are characterized by excessive mucus production, which not only obstructs the airways but also creates an environment conducive to bacterial infections. This accumulation of mucus can further exacerbate inflammation and airway hyperreactivity, creating a vicious cycle that complicates disease management and negatively impacts the quality of life for affected individuals. Addressing mucin overproduction through targeted therapies may help alleviate these symptoms and improve overall respiratory function.

- Mucins: Mucin is a type of glycoprotein that is a key component of mucus, a viscous secretion found in various tissues throughout the body. Mucins play a crucial role in protecting and lubricating the surfaces of the respiratory, gastrointestinal, and urogenital tracts, as well as other mucosal surfaces. Mucins, primarily MUC5AC, make mucus more viscous. While mucus is essential for trapping pathogens, overproduction can lead to airway obstruction. Regulating mucin expression is vital for managing diseases characterized by heightened mucus production.

- Extracellular Matrix (ECM) Remodeling: ECM remodeling refers to the dynamic process by which the composition and structure of the ECM are modified in response to physiological and pathological stimuli. The ECM is a complex network of proteins and carbohydrates that provides structural and biochemical support to surrounding cells. Key components include collagen, elastin, fibronectin, and glycoproteins. Pathways involving transforming growth factor-beta (TGF-beta) are key players in ECM remodeling, which contributes to airway hyperresponsiveness and remodeling in conditions like asthma and pulmonary fibrosis. Targeting these pathways may mitigate disease progression.

Essential oils that work on this pathway: Bergamot, cajeput (*Melaleuca cajuputi*), copaiba, eucalyptus (*E. globulus*), hinoki, peppermint, spike lavender (*Lavandula latifolia*), and tangerine

Muscarinic and Adrenergic Pathways— The autonomic nervous system plays a significant role in airway tone and reactivity. Muscarinic pathways are part of the parasympathetic nervous system and involve muscarinic acetylcholine receptors (mAChRs), which are G protein-coupled receptors that respond to the neurotransmitter acetylcholine. These receptors are found in various tissues throughout the body, including the heart, smooth muscle, and glands. They serve to slow heart rate and reduce contractility, increase smooth muscle contraction (enhancing digestion and gut motility), and stimulate secretions from salivary, lacrimal, and respiratory glands. Adrenergic pathways are part of the sympathetic nervous system and involve adrenergic receptors that respond to catecholamines, primarily norepinephrine and epinephrine (adrenaline). These receptors are divided into two major classes: alpha (α) and beta (β) adrenergic receptors, each with subtypes. Adrenergic receptors increase heart contractility and contractility (β1 receptors), and cause vasoconstriction (α1 receptors) or vasodilation (β2 receptors), depending on the receptor type and location; are involved in metabolism; and relax bronchial smooth muscle (β2 receptors), facilitating bronchodilation. Both muscarinic and adrenergic pathways are crucial for homeostasis and the body's response to stress, regulating a variety of physiological functions.

- Muscarinic antagonists: Molecules that bind to a receptor to block or dampen its biological response are called antagonists. Muscarinic antagonists block M3 muscarinic receptors, reducing bronchoconstriction and mucus secretion in conditions like COPD and asthma.
- β-Adrenergic agonists: Molecules that bind to a receptor and activate it to trigger a biological response are called agonists. Activation of β2-adrenergic receptors leads to bronchodilation, making short-acting and long-acting β2 agonists (SABAs and LABAs) essential agents for managing asthma and COPD symptoms.

Essential oils that work on this pathway: Ajowan (*Trachyspermum ammi*), black cumin (*Nigella sativa*), ginger, melissa, and vetiver

Immune System Modulation—The immune response plays a crucial role in maintaining respiratory health, as it involves a complex interplay of various immune cells that work together to protect the airways from pathogens and irritants. Key players in this process include macrophages, neutrophils, and lymphocytes, which not only identify and eliminate harmful invaders but also release signaling molecules that modulate inflammation. This intricate network of interactions can either promote healing and restore normal function or contribute to airway inflammation and respiratory diseases, such as asthma and chronic obstructive pulmonary disease (COPD). Understanding these immune mechanisms is essential for developing targeted therapies to enhance respiratory health and manage inflammatory conditions effectively.

- T helper cells: The balance between Th1 and Th2 immune responses is crucial for respiratory health. An overactive Th2 response can lead to allergic asthma, while a robust Th1 response is generally protective against infections. Targeting this balance may help manage asthma effectively.
- Dendritic cells and macrophages: Dendritic cells and macrophages are essential for recognizing allergens and pathogens in the airway, acting as the first line of defense in the immune response. These antigen-presenting cells capture and process foreign substances, subsequently activating T helper cells, particularly the Th2 subset, which is pivotal in orchestrating allergic responses and inflammation. Modulating their activity can influence Th2-mediated

inflammation and may improve respiratory outcomes by reducing excessive inflammatory responses and promoting a balanced immune reaction. By targeting the signaling pathways and interactions of these immune cells, therapeutic strategies can be developed to alleviate symptoms of allergic diseases and enhance overall respiratory health, potentially leading to better management of conditions like asthma and allergic rhinitis.

Essential oils that work on this pathway: Hinoki leaf, frankincense

Vascular Endothelial Growth Factor (VEGF) Pathway—VEGF serves a critical role in mediating angiogenesis, which is the biological process through which new blood vessels form from existing ones. This process is essential for growth, development, and wound healing, as it ensures that tissues receive adequate blood supply and nutrients. VEGF is not only vital for normal physiological processes but also plays a significant role in various pathological conditions, including cancer and chronic inflammatory conditions. In the context of respiratory diseases, VEGF is implicated in airway remodeling, a process characterized by structural changes in the airways that can lead to increased vascularization, thickening of the airway walls, and heightened sensitivity to allergens. Controlling VEGF signaling could be particularly beneficial in conditions characterized by excessive vascularization, such as asthma, where the overproduction of blood vessels contributes

to airway obstruction and inflammation. By targeting VEGF pathways, therapeutic interventions may help restore normal vascular function and mitigate the adverse effects of airway remodeling, ultimately improving respiratory health and quality of life for affected individuals.

Essential oils that work on this pathway: Lemon and ylang ylang

Endocannabinoid System—The endocannabinoid system, which includes cannabinoid receptors CB1 and CB2, has emerged as a potential target for managing respiratory health due to its complex role in modulating various physiological processes. CB1 receptors are primarily found in the central nervous system and play a significant role in regulating neurotransmitter release, while CB2 receptors are predominantly located in the immune system and peripheral tissues, including the lungs. This distribution suggests that the endocannabinoid system can influence both the nervous and immune responses in the respiratory tract. Research has shown that activation of these receptors can help reduce inflammation, alleviate bronchoconstriction, and promote airway relaxation, making them promising candidates for treating respiratory conditions such as asthma and chronic obstructive pulmonary disease (COPD). Additionally, the endocannabinoid system is involved in pain modulation and the regulation of mucus production, further highlighting its potential in addressing multiple aspects of respiratory health. By exploring cannabinoid-based therapies, researchers aim to develop innovative treatments that harness the therapeutic benefits of the endocannabinoid system while minimizing side effects, ultimately improving the management of respiratory diseases and enhancing patients' quality of life.

Essential oils that work on this pathway: Black pepper (beta-caryophyllene), cananga (beta-caryophyllene), copaiba (beta-caryophyllene), hemp (selina-3,7(11)-diene), melissa (beta-caryophyllene, ylang ylang (beta-caryophyllene)

Neurotransmitter Pathways— Neurotransmitters, such as substance P and neurokinins, play a significant role in regulating airway smooth muscle contraction and inflammation, making them critical players in respiratory physiology. Substance P, a neuropeptide, is released from sensory nerves in the airways and is known to induce bronchoconstriction, leading to narrowing of the air passages, which can exacerbate conditions like asthma and COPD. Neurokinins, including neurokinin A and neurokinin B, also contribute to airway hyperreactivity and inflammation by promoting the release of proinflammatory mediators from immune cells, further complicating respiratory conditions. Modulating these neurotransmitter pathways holds promise for improving symptoms in respiratory diseases by potentially reducing excessive airway constriction and inflammation. Therapeutic strategies that target substance P and neurokinin receptors could lead to the development of novel treatments that

alleviate bronchoconstriction, decrease mucus production, and enhance overall airway function. By addressing the neurogenic components of respiratory diseases, such interventions may provide a more comprehensive approach to managing symptoms and improving the quality of life for individuals suffering from chronic respiratory conditions.

Essential oils that work on this pathway: Black sage (*Cordia verbenacea*), linalool

Airway Epithelial Integrity—Maintaining the integrity of the airway epithelium is vital for respiratory health, as this epithelial layer serves as the first line of defense against environmental insults, including pathogens, allergens, and pollutants. The airway epithelium is composed of specialized cells that not only provide a physical barrier but also play an active role in immune responses and mucociliary clearance. This barrier is essential for preventing the entry of harmful substances into the underlying tissues, thereby reducing the risk of infections and inflammatory responses that can lead to chronic respiratory diseases.

When the integrity of the airway epithelium is compromised, as seen in conditions like asthma, chronic bronchitis, or exposure to tobacco smoke, it can result in increased permeability and inflammation. This disruption can lead to a cascade of pathological changes, including airway hyperresponsiveness, mucus overproduction, and remodeling of the airway structure. Furthermore, a damaged epithelium can impair the function of cilia, which are responsible for clearing mucus and debris from the airways, leading to further accumulation of harmful substances and exacerbating respiratory symptoms.

Therefore, strategies aimed at preserving or restoring the integrity of the airway epithelium are crucial for promoting respiratory health. This can involve the use of anti-inflammatory agents, antioxidants, and therapies that enhance epithelial repair and regeneration. By ensuring the airway epithelium remains intact and functional, it is possible to improve lung function, reduce the frequency of exacerbations, and enhance the overall quality of life for individuals with respiratory conditions.

Essential oils that work on this pathway: Mastic leaf (*Pistacia lentiscus*), peppermint

Pulmonary Surfactant Regulation— Surfactant proteins (SP-A, SP-B, SP-C, and SP-D) are essential for reducing surface tension in the alveoli, which are tiny, balloon-like structures at the end of the respiratory bronchioles in the lungs that are vital for gas exchange. In addition to their role in maintaining alveolar stability, these proteins are also important for pulmonary host defense, helping to protect the lungs from infections and inflammation. Enhancing the function of surfactant proteins may provide therapeutic benefits for conditions characterized by surfactant deficiency or dysfunction, such as neonatal respiratory distress syndrome and acute respiratory distress syndrome (ARDS). By improving surfactant function, it may be possible to restore normal lung mechanics, enhance gas exchange, and support overall respiratory health in affected individuals.

Essential oils that work on this pathway: Currently unknown.

The interplay of various Molecular or Cellular Targets/Pathways underscores the complexity of respiratory health. Advances in research continue to illuminate new targets and therapeutic opportunities, providing hope for effective management and prevention of respiratory diseases. By understanding and modulating these fundamental biological processes, we can foster better respiratory health outcomes and enhance the quality of life for individuals facing these challenging conditions.

Hormonal and Endocrine Effects— Certain essential oils can influence endocrine function and hormone regulation. For example, clary sage oil is often cited for its potential benefits in balancing hormones and alleviating symptoms related to menstrual irregularities. This may be attributed to its sclareol content, which is a sesquiterpene alcohol with a similar structure to estradiol. Impressively, some essential oils can improve hormone profiles simply from inhalation, which is a testament to the power of essential oils. The diverse mechanisms by which they exert their effects reveal their multifaceted therapeutic potential. By understanding these pathways, practitioners and individuals can more effectively utilize essential oils for a wide range of health concerns.

The endocrine system plays a vital role in maintaining homeostasis through hormone production and regulation. This complex system involves a range of molecular and cellular targets, pathways, and mechanisms that govern endocrine function. Understanding these components is essential for elucidating how hormones influence physiological processes and how disruptions can lead to various disorders.

Hormone Receptors: Hormones exert their effects by binding to specific receptors on target cells. These receptors can be classified into:

- Cell Surface Receptors: Typically for peptide and catecholamine hormones (e.g., adrenaline, noradrenaline, and dopamine). These receptors are often G protein-coupled receptors (GPCRs) or receptor tyrosine kinases (RTKs) that activate intracellular signaling cascades upon hormone binding.
- Intracellular Receptors: Found in the cytoplasm or nucleus, these receptors bind to steroids (e.g., testosterone, estradiol, progesterone) and thyroid hormones (e.g., triiodothyronine, thyroxine, and calcitonin). Once bound, they typically act as transcription factors, regulating gene expression and protein synthesis.

Second Messengers: Upon hormone-receptor interaction, intracellular second messengers are often activated, amplifying the signal within the cell. Common second messengers include:

- Cyclic Adenosine Monophosphate (cAMP): Often involved in GPCR pathways, regulating various cellular responses including metabolism and gene expression. It translates hormonal signals into various cellular responses through its ability to activate protein kinase A.
- Inositol Triphosphate (IP3) and Diacylglycerol (DAG): Generated by phospholipase C activation, leading to calcium release and protein kinase C activation.

Target Cell Types: Different hormones target specific cell types, which have the necessary receptors for the hormone's action. For example:

- Adipocytes: Respond to insulin and leptin, regulating glucose metabolism and energy homeostasis.
- Muscle cells: Particularly responsive to insulin for glucose uptake and metabolism.
- Endothelial cells: Targeted by nitric oxide and hormones like angiotensin II, regulating vascular tone and blood pressure.

Feedback Mechanisms: Endocrine functions are tightly regulated by feedback loops, often negative feedback, where the effects of a hormone inhibit its further release. For example, elevated thyroid hormones discourage further secretion from the hypothalamus and pituitary gland.

Hypothalamic-Pituitary Axis: A core component of endocrine regulation, the hypothalamus releases triggering or inhibiting hormones that act on the anterior pituitary gland, which subsequently produces tropic hormones that stimulate other endocrine glands (e.g., adrenal, thyroid, gonadal).

Signal Transduction Pathways: Hormonal signaling involves complex pathways that transduce—the process of converting or transforming an external signal into a specific intracellular response—extracellular signals, which is critical for cellular communication and responses to both other cells and their environment. Key pathways include:

- MAPK/ERK pathway: Activated by growth factors and certain hormones, regulating cell proliferation and differentiation.

- PI3K/Akt pathway: Important for glucose metabolism and cell survival, often triggered by insulin and IGF (insulin-like growth factor).

- Glycogenolysis and gluconeogenesis: Hormones like glucagon and cortisol stimulate these metabolic processes, crucial for maintaining blood glucose levels during fasting or stress.

Calcium Signaling: Hormones such as parathyroid hormone (PTH) regulate calcium levels through mechanisms involving bone resorption, renal reabsorption, and intestinal absorption.

Hormonal Regulation of Gene Expression: Steroid hormones, upon binding to their intracellular receptors, facilitate transcription of target genes that mediate various biological processes, including metabolism, immune response, and reproductive functions.

Highlighting the potential to regulate endocrine function and promote hormone balance through various methods of use, essential oils have shown potential to do so topically, aromatically, and orally. Topical application of blue spruce (*Picea pungens*)

increased total and free testosterone by an average of 9% and 26.8% respectively.[49] A preclinical model found that oral administration of black cumin essential oil restored total testosterone and sex-hormone-binding globulin levels and significantly increased free testosterone and free androgen index.[50]

Several essential oils, or their constituents, also interact with hormonal targets like MAPK/ERK and PI3k/Akt:

- Agarwood (*Aquilaria* spp.)—PI3k[51]
- *ar*-Turmerone—MAPK[52]
- Bay laurel (*Laurus nobilis*)—PI3k[53]
- Black pepper—MAPK[54]
- Copaiba—MAPK, PI3k, Akt[55]
- Citron (*citrus medica*)—ERK
- German chamomile CT chamazulene—MAPK, PI3K, AKt[56]
- Lavender—MAPK[57]
- Orange (sweet)[58]
- Tea tree—MAPK[59]
- Thyme—PI3k, Akt[60]
- Trans-Cinnamaldehyde—MAPK[61]
- Ylang ylang—MAPK, PI3K, Akt[62]

Other essential oils can improve glycogenesis:

- Myrtle (*Myrtus communis*)[63]

Or calcium channels/signaling:

- Bergamot[64]
- Black cumin[65]
- Citronella[66]
- Fennel[67]
- Lavender[68]
- Melissa CT citral/citronellal[69]
- Neroli[70]

- Rosemary CT 1,8-cineole.[71]
- Spearmint.[72]
- Vetiver.[73]

The endocrine system is characterized by a complex interplay of molecular and cellular targets, sophisticated signaling pathways, and regulatory mechanisms that ensure precise control of physiological functions. A comprehensive understanding of these elements is critical for advancing therapeutic strategies involving essential oils that target endocrine disorders. By comprehending how these pathways operate, researchers and clinicians can better assess and regulate hormone-related functions in health and disease. Inhalation of rose or geranium essential oil increased salivary estrogen levels in premenopausal women.[74] Thus, harnessing the therapeutic potential of essential oils offers promising pathways to naturally modulate hormonal balance and enhance endocrine health, paving the way for innovative alternatives in the management of hormonal disorders.

Gene Expression and DNA Methylation—Essential oils contain bioactive compounds that can modulate gene expression by interacting with transcription factors, nuclear receptors, and epigenetic regulators. These compounds can upregulate or downregulate specific genes involved in inflammation, oxidative stress, metabolism, immune function, sleep, and much more. For instance:

- β-Caryophyllene (found in copaiba and black pepper essential oils) interacts with cannabinoid receptor type 2 (CB2), modulating genes related to inflammation, the stress response, and immune responses.[75, 76, 77]

- Eugenol (from clove oil) influences NF-κB signaling, affecting genes involved in inflammation and cellular protection.[78, 79]

DNA methylation is a critical epigenetic modification that controls gene activity by adding methyl groups to cytosine residues, typically silencing gene expression. Essential oils can influence DNA methylation patterns by modulating the activity of DNA methyltransferases (DNMTs) and histone deacetylases (HDACs), leading to epigenetic reprogramming.

- Patchouli essential oil downregulates a family of HDACs—HDAC4, HDAC5, HDAC7, HDAC8, and HDAC9—related to various types of cancer and increases the expression of peptide tyrosine-tyrosine (PYY), which is a gut hormone related to inflammation control and anticancer activity.[80]

- The ability of monoterpenes—like camphor and 1,8-cineole—to modulate epigenetic marks (such as DNA methylation or histone modifications) allows them to stimulate error-free DNA repair pathways, particularly nucleotide excision repair (NER), which corrects DNA damage from UV exposure and oxidative stress.[81]

- (E)-Nerolidol, a sesquiterpene alcohol found in essential oils like niaouli (*Melaleuca quinquenervia*) and neroli, reduces the overexpression of DNA methyltransferase 1, reducing inflammation and cell activation to decrease depression.[82]

Several molecular pathways mediate the effects of essential oils on gene expression and DNA methylation:

Nrf2/ARE Pathway: Many essential oil components, such as limonene and eugenol, activate the Nrf2 pathway, enhancing antioxidant defense mechanisms by upregulating genes like NQO1 and HO-1.[83, 84]

NF-κB Signaling: Essential oils like fennel and lavender modulate NF-κB activity, downregulating proinflammatory cytokine genes such as IL-6 and TNF-α.[85, 86]

PI3K/Akt/mTOR Pathway: Some essential oils influence cellular growth and longevity by interacting with the PI3K/Akt/mTOR axis, impacting genes involved in metabolism and cellular repair. Copaiba is one of these, with research demonstrating its ability to regulate this important pathway in brain cells.[87]

SIRT1 Activation: SIRT1 is a key protein involved in cellular defense and aging. Interestingly, 1,8-cineole activates this and the Nrf2 pathway to combat oxidative stress and relieve mitochondrial dysfunction associated with poor glucose metabolism.[88]

Emerging clinical research supports this preclinical research, demonstrating essential oils have widespread and dramatic effects on gene expression. From lavender improving gene expression related to sleep and circadian rhythm genes to frankincense's impact on inflammation, it is abundantly clear that essential oils fine-tune gene expression so your body can do its intended job more efficiently.[89, 90] The end result is a healthier and more vibrant you.

Essential oils exert profound effects on gene expression and DNA methylation through interactions with key molecular pathways, transcription factors, and epigenetic regulators. These effects contribute to their therapeutic potential in inflammation modulation, neuroprotection, detoxification, metabolic balance, circadian rhythm, and much more.

Identifying specific molecular targets for essential oils is pivotal for understanding their effectiveness and optimizing their therapeutic applications. The complexity of essential oils means that they can interact with numerous targets and pathways, allowing for their application across several health conditions. The therapeutic potential of essential oils extends to various healthcare settings, including aromatherapy, integrative medicine, and dermatology. Understanding their mechanisms and molecular targets allows for:

Personalized Therapeutic Approaches: By matching essential oils to individual health needs—considering aspects such as body weight, sex, and specific health conditions—healthcare practitioners can create tailored therapeutic regimens.

Improvements in Safety: Recognizing the potential for interactions among essential oil components and with conventional medications is crucial. Knowledge of molecular targets, pathways, and mechanisms reveals intersections with other therapeutics and facilitates the safe use of essential oils while minimizing adverse effects.

Enhanced Outcomes in Chronic Conditions: Chronic illnesses often involve multifactorial mechanisms. The polypharmacological nature of essential oils makes them advantageous in managing conditions such as chronic pain, mental health disorders, and inflammatory diseases, offering more than symptomatic relief and potentially altering disease progression.

Wound Healing and Skin Health: The antimicrobial and anti-inflammatory properties of essential oils like lavender and tea tree oil lend themselves to applications in dermatological care, promoting healing and preventing infections.

Support for Mental Health: The anxiolytic and antidepressant effects of various essential oils have garnered attention in psychiatric and psychological treatments. Their integration into traditional therapies, such as cognitive-behavioral therapy, can enhance treatment outcomes for anxiety and depression.

While essential oils possess considerable therapeutic potential, several challenges must be addressed to optimize their use:

Standardization and Quality Control: The composition of essential oils can vary significantly among different sources and batches. Ensuring consistent quality through rigorous testing and standardization is vital for their effective therapeutic use.

Dosage Determination: Establishing optimal dosing regimens is challenging due to the variable concentrations of bioactive components in essential oils. The current lack of pharmacokinetic and pharmacodynamic data for most essential oils limits dosing determination. Extrapolation from animal data is not always ideal, nor applicable, to human efficacy. Research into dose-response relationships and comparative studies with conventional therapies can help define appropriate dosages.

Interactions with Conventional Medications: While interactions with pharmaceuticals are rarely observed, the potential for essential oils to interact with them is possible, potentially amplifying or diminishing their effects. Comprehensive understanding of these interactions is necessary to devise safe treatment plans.

Essential oils represent a remarkable resource in natural therapeutics, shaped by their inherent polypharmacology and complex mixture of bioactive compounds. By understanding their molecular targets, mechanisms, and pathways, we can maximize their therapeutic potential and improve health outcomes. Through further research and enhanced integration of essential oils into clinical practice, we stand on the brink of realizing their full promise as holistic and effective natural therapeutics. As we continue to explore and appreciate the rich complexity of these substances, their role in integrative health care will undoubtedly expand, offering new avenues for promoting wellness and treating diverse health conditions.

Molecular Targets, Biological Pathways, and Mechanisms for Key Essential Oils

The molecular and cellular targets, biological pathways, and mechanisms discussed in this chapter were compiled using the author's personal research database and his books *Medicinal Essential Oils: The Science and Practice of Evidence-based Essential Oil Therapy* and *Co2 Aromatics, Medicinal Herbs, and Targeted Nutraceuticals for Healing and Greater Wellness*. Citations, references, and additional details are available within these books. This work includes profiles for forty-seven of the most commonly used essential oils (sixty plus botanical species), with suggested therapeutic applications based on the identified targets, pathways, and mechanisms. Detailed protocols and safety guidelines can be found in the author's other publications.

A molecular target refers to any molecule involved in a specific biological process. Molecular targets can include proteins (like enzymes and receptors), nucleic acids (like DNA or RNA), lipids, or carbohydrates. Essentially, any molecule that a therapeutic agent interacts with at the molecular level can be considered a molecular target.

A cellular target refers to entire cells, specific cellular processes, or any part of a cell that a therapeutic agent influences. This can include receptors, enzymes, ion channels, or any other cellular component that affects cellular function.

A biological pathway is a series of interrelated biochemical reactions or molecular interactions that lead to a specific cellular outcome or function. These pathways play essential roles in various biological processes, including metabolism, signal transduction, gene regulation, and cellular communication.

In the context of pharmacology and therapeutic agents, a mechanism refers to the specific biochemical or physiological processes through which a therapeutic agent exerts its effects on the body. The mechanism encompasses the interactions of the agent with biological molecules (such as proteins, enzymes, or receptors) and the subsequent cascade of events that lead to a therapeutic outcome.

This chapter provides a comprehensive overview of the molecular pharmacology of essential oils, illuminating the intricate web of targets, pathways, and mechanisms that underlie their therapeutic effects. By integrating information from this author's extensive research database and key texts in the field, we gain valuable insights into how these natural complex substances interact with biological systems at both the molecular and cellular levels. The exploration of forty-seven commonly used essential oils, along with their suggested therapeutic applications, empowers practitioners and researchers alike to harness the potential of these oils in evidence-based therapeutic practices. As we deepen our understanding of the complexities surrounding how essential oils impact health and wellness, refine dosing and methodologies, and assess their safety, these oils are increasingly emerging as leading natural solutions for various human ailments. This growing body of knowledge not only furthers our understanding of essential oils but also underscores their relevance in a holistic approach to health care.

BASIL (SWEET/TROPICAL)

Basil (*Ocimum basilicum*), often celebrated for its culinary applications, has emerged as a powerhouse of therapeutic potential with a range of bioactive compounds that demonstrate significant anticancer,

neuroprotective, analgesic, antimicrobial, cardiovascular, and anti-aging effects. The essential oil extracted from basil contains key constituents such as linalool, estragole, eugenol, and 1,8-cineole, which collectively interact with various molecular targets to exert their therapeutic actions. Notably, basil exhibits anticancer activity against glioblastoma, leukemia, and other cancer cell lines by inducing apoptosis, modulating cell cycle regulatory proteins, and inhibiting oncogenic signaling pathways like PI3K/Akt and NF-κB. Simultaneously, basil's neuroprotective effects are facilitated through the modulation of serotonergic and cortisol pathways, leading to increased serotonin levels and reduced cortisol, reminiscent of conventional antidepressant mechanisms. Additionally, basil showcases potent analgesic properties by influencing central pain sensitization pathways and inhibiting inflammatory mediators. Its antimicrobial efficacy targets bacterial membranes and biofilm formation, making it effective against a spectrum of pathogens. Furthermore, basil supports cardiovascular health by regulating lipid metabolism and reducing oxidative stress, while its antioxidant properties contribute to cellular longevity and anti-aging benefits. As research continues to illuminate the myriad effects of basil, it stands out as a multifaceted natural remedy with promising applications across various health domains.

SUGGESTED THERAPEUTIC USE:

Cancer: Glioblastoma, leukemia, various other cancer types due to its anticancer activity and apoptosis induction

Neurological Disorders: Depression and anxiety, cognitive impairments and memory issues (such as those related to Alzheimer's or neurological degeneration)

Pain: Chronic pain conditions, inflammatory pain

Infections: Bacterial infections, fungal infections, parasitic infections

Cardiovascular Health: High cholesterol or triglyceride levels, atherosclerosis prevention

Respiratory Issues: Asthma and allergies, acute ear infections

Insect Bites or Infestations: Pest control for home environments

Skin Conditions: Acne and other dermatological issues, signs of premature skin aging

Overall Well-being: General stress relief and improved mood, support for immune system function

Cellular Aging: Conditions related to accelerated aging at the cellular level (atherosclerosis, Alzheimer's disease, Parkinson's disease. Metabolic syndrome, diabetes, osteoporosis, cancer, skin aging, chronic inflammatory conditions, liver fibrosis, kidney disease)

MOLECULAR AND CELLULAR TARGETS, BIOLOGICAL PATHWAYS, AND MECHANISMS:

1. Anticancer Activity

Molecular or Cellular Target/Pathway: Glioblastoma, leukemia, and various cancer cell lines.

Mechanisms: Induction of apoptosis (programmed cell death) in cancer cells. Anti-proliferative effects by modulating cell cycle regulatory proteins. Protection of DNA from damage and mutations (suggesting an influence on DNA repair mechanisms and oxidative stress pathways). Likely involvement of antioxidant enzymes (e.g., superoxide dismutase, catalase) reducing oxidative stress-induced carcinogenesis. Potential inhibition of oncogenic signaling pathways such as PI3K/AKT, NF-κB, and MAPK.

2. Neurological Effects (Antidepressant, Cognitive, and Stress Relief)

Molecular or Cellular Target/Pathway: Serotonergic—anything related to serotonin (a neurotransmitter that plays a key role in regulating mood, emotion, sleep, and other functions in the body), which can describe neurons that release serotonin, receptors that bind to serotonin, or pathways influenced by serotonin—and cortisol pathways.

Mechanisms: Increases serotonin (5-HT) levels while reducing cortisol, similar to selective serotonin reuptake inhibitors (SSRIs), suggesting an interaction with monoamine neurotransmitter systems. Modifies neuronal activity pathways, as shown by altered FOS protein expression, which is involved in central nervous system plasticity and pain processing. Protein expression refers to the production of FOS proteins, which are a family of immediate early gene products involved in cellular signaling and gene regulation. These proteins, particularly c-FOS, are rapidly induced in response to various stimuli, including growth factors, stress, and neuronal activity. FOS proteins can form part of the AP-1 transcription factor complex, which is crucial for processes such as cell proliferation, differentiation, and apoptosis. The measurement of FOS protein expression is often used as a marker for cellular activation and is commonly studied in neuroscience to understand neuronal responses to stimuli. Enhances blood flow to the prefrontal cortex, potentially via vasodilation mediated by nitric oxide (NO) or prostaglandins. Alleviates memory impairment and hippocampal neurodegeneration, possibly through neuroprotective antioxidant effects. Hippocampal neurodegeneration involves the loss or damage of neurons in the hippocampus, a brain region crucial for memory and learning. This process can lead to cognitive decline and is commonly associated with conditions like Alzheimer's disease, stress, and other neurological disorders.

3. Analgesic (Pain-Relieving) Properties

Molecular or Cellular Target/Pathway: Pain pathways in the CNS and inflammatory mediators.

Mechanisms: Modifies FOS protein expression, impacting central pain sensitization. Inhibits prostaglandin and prostacyclin synthesis, reducing inflammation and nociceptive signaling. Prostaglandins are lipid compounds derived from fatty acids, particularly arachidonic acid, and are produced by nearly all tissues in the body. They play critical roles in various physiological processes, including inflammation, pain

sensation, regulation of blood flow, and modulation of immune responses. Prostacyclin, also known as prostaglandin I2 (PGI2), is a specific type of prostaglandin that is particularly important for vascular function. It is synthesized primarily in the endothelium (the inner lining of blood vessels) and acts as a potent vasodilator, helping to regulate blood pressure and inhibit platelet aggregation. Enhances bioavailability and duration of pain relief when combined with beta-cyclodextrin. Likely interacts with opioid receptors or TRPV1 (Transient Receptor Potential Vanilloid 1) involved in pain perception.

4. Antimicrobial, Antifungal, and Antiparasitic Activity

Molecular or Cellular Target/Pathway: Bacterial cell walls, biofilm formation, and microbial metabolism.

Mechanisms: Disrupts bacterial membranes in Gram-negative and Gram-positive bacteria (e.g., *E. coli*, *K. pneumoniae*, *S. aureus*, *P. aeruginosa*). Inhibits biofilm—structured communities of bacteria that adhere to surfaces and are embedded in a self-produced extracellular matrix composed of polysaccharides, proteins, and nucleic acids—formation in *P. aeruginosa* and *S. aureus*, preventing antibiotic resistance mechanisms. Exhibits fungistatic and fungicidal activity against *Candida* spp. by interfering with ergosterol synthesis—a vital process fungi use to create a component of their cell membranes—or fungal respiration—the process by which fungi convert nutrients into energy. Inhibits mycotoxin production by *Aspergillus* species, suggesting modulation of fungal metabolic pathways. Effective against parasitic infections (e.g., *Giardia lamblia, Acanthamoeba* spp.), possibly by disrupting parasite membrane integrity.

5. Cardiovascular Benefits

Molecular or Cellular Target/Pathway: Lipid metabolism, cholesterol regulation, and endothelial function.

Mechanisms: Reduces total cholesterol, LDL, and triglycerides while increasing HDL, likely by influencing PPAR (Peroxisome Proliferator-Activated Receptors). PPARs are a group of nuclear receptor proteins that play key roles in regulating lipid metabolism, including cholesterol homeostasis. Prevents atherosclerosis progression by reducing lipid accumulation in the aorta. Protects myocardium from ischemic damage (insufficient blood supply to a particular tissue or organ), possibly by reducing oxidative stress and inflammation.

6. Respiratory and Immune-Modulating Effects

Molecular or Cellular Target/Pathway: Immune cells, cytokines, and inflammatory mediators.

Mechanisms: Reduces acute ear infections, indicating antibacterial and anti-inflammatory effects. Inhibits inflammatory responses related to asthma and allergies, likely via histamine modulation and suppression of proinflammatory cytokines (e.g., IL-6, TNF-α). Enhances immune response by stimulating natural killer (NK) cells or modulating adaptive immunity.

7. Insecticidal and Pest-Repellent Activity

Molecular or Cellular Target/Pathway: Insect neuroreceptors and enzyme systems.

Mechanisms: Effective against dust mites, mosquito larvae, stored grain insects, likely by interfering with AChE activity, leading to neurotoxicity in insects. Synergistic effects with diatomaceous earth, suggesting interaction with physical and biochemical insecticidal mechanisms.

8. Dermatological and Cosmetic Applications

Molecular or Cellular Target/Pathway: Keratinocytes, sebaceous glands, and microbial flora.

Mechanisms: Antiseptic and keratolytic effects improve acne via disruption of bacterial biofilms and sebum regulation. A keratolytic effect means that the substance aids in the removal or shedding of the outer layer of skin composed of dead skin cells. Increases skin temperature and sensory stimulation, suggesting modulation of cutaneous blood flow. Antioxidant activity prevents premature skin aging by protecting against oxidative stress.

9. Cellular Longevity and Anti-Aging Effects

Molecular or Cellular Target/Pathway: Telomeres, oxidative stress pathways.

Mechanisms: Prevents telomere shortening after oxidative stress, which suggests involvement in DNA repair and longevity pathways. Telomeres can be thought of as the plastic tips (aglets) found at the ends of shoelaces. Just like aglets protect the ends of shoelaces from fraying and unraveling, telomeres protect the ends of chromosomes from degradation. They prevent the DNA strands from becoming damaged during the process of cell division. Over time, if the shoelaces are used frequently without care, the aglets can become worn down or may break off, leading to fraying shoelaces that are harder to manage. This parallels how telomeres shorten with each cell division. As cells replicate, telomeres get shorter, just as the protective ends of shoelaces can wear out over time. Increases the length of already shortened telomeres, indicating potential telomerase activation or enhanced cellular repair mechanisms. In essence, this is turning back the biological clock of your cells, restoring them to more youthful functions.

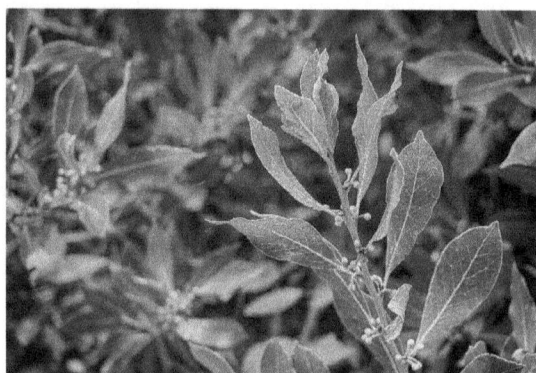

BAY LAUREL

Bay laurel (*Laurus nobilis*) essential oil, rich in 1,8-cineole, linalool, and other terpenes, demonstrates a diverse array of therapeutic properties. Research spans *in vitro*, *in vivo*, and *in silico* studies, highlighting its potential in cancer therapy, neurological disorders, microbial infections, and metabolic regulation. Its complex chemical profile allows for multifaceted interactions with biological systems, offering promise for various health applications. Notably, bay laurel exhibits potent antimicrobial activity against a broad spectrum of bacteria and fungi, surpassing some conventional treatments in certain cases. Moreover, studies suggest its efficacy in combating cancer cell proliferation, modulating neurological functions, and managing metabolic conditions like diabetes and obesity. The oil's ability to inhibit key enzymes and interact with cellular receptors underscores its therapeutic versatility and makes it a valuable natural solution for a variety of ailments.

SUGGESTED THERAPEUTIC USE:

Neurological Conditions: Alzheimer's disease, dementia, myasthenia gravis, Lewy body dementia, schizophrenia, cognitive decline, memory, attention

Cancer: Myeloid leukemia, breast, kidney, neuroblastoma

Infections: Bacterial, antibiotic-resistant, fungal, viral, mycotoxin reduction

Metabolic Disorders: Type 2 diabetes, obesity, high cholesterol or triglycerides, glycation-related complications (diabetic retinopathy, diabetic nephropathy, diabetic neuropathy, cardiovascular disease, neurodegenerative diseases, kidney disease, osteoporosis, aging)

Inflammatory Conditions: Arthritis (rheumatoid arthritis, osteoarthritis), asthma, IBD

Cardiovascular Conditions: Cardiovascular disease, high blood pressure, atherosclerosis

Pain Management: General pain management, inflammatory pain, neuropathic pain

Organ and Tissue Protection: Oxidative stress (cancer, IBD, diabetes, neurodegenerative conditions, cardiovascular disease, liver or kidney disease, metabolic syndrome, fibromyalgia, chronic fatigue syndrome, skin aging), cellular damage

Kidney Issues: Kidney disease

Pest Problems: Tick and mite infestations

MOLECULAR AND CELLULAR TARGETS, BIOLOGICAL PATHWAYS, AND MECHANISMS:

1. Anticancer (Myeloid Leukemia, Breast Cancer, Melanoma, Kidney Cancer, Neuroblastoma)

Molecular or Cellular Target/Pathway: PI3K/mTOR pathway, cell cycle regulation, apoptosis induction.

Mechanisms: Bay laurel essential oil inhibits PI3K/mTOR signaling, leading to decreased cancer cell proliferation. The oil induces apoptosis through interactions with cellular signaling pathways. It also inhibits cell cycle progression. Lastly, its constituents—methyleugenol, sabinene, 1,8-cineole, and alpha-terpinyl acetate—interact with cancer-related targets.

2. Analgesic and Anti-inflammatory

Molecular or Cellular Target/Pathway: Arachidonic acid cascade, soluble epoxide hydrolase (sEH).

Mechanisms: Bay laurel oil inhibits sEH, reducing the production of proinflammatory mediators. It also modulates pain pathways, potentially through interactions with opioid receptors. Its constituents—methyleugenol, linalool, terpinen-4-ol, and eugenol—may also interact with sEH. The arachidonic acid cascade and sEH are key players in inflammatory processes. Inhibiting sEH is important because this enzyme breaks down naturally occurring compounds called epoxides that help reduce pain and inflammation in the body. When sEH is active, it converts these protective epoxides into other substances that promote inflammation, leading to increased pain and discomfort. By blocking sEH, bay laurel oil can increase the levels of these beneficial epoxides, which can help reduce inflammation and alleviate pain, making it a useful target for pain relief strategies.

3. Cognitive Enhancement (Alzheimer's, Myasthenia Gravis, Lewy Body Dementia, Schizophrenia)

Molecular or Cellular Target/Pathway: AChE, sEH, oxidative stress pathways.

Mechanisms: Inhibition of AChE by bay laurel oil increases acetylcholine levels to improve memory and cognition. The oil's activity is believed to be mediated by interaction of 1,8-cineole and alpha-terpinyl acetate with acetylcholinesterase. By reducing harmful inflammatory responses—inhibition of sEH—in the brain, this approach could potentially slow down the progression of diseases like Alzheimer's, offering a promising avenue for improving overall health and quality of life in those affected by these conditions. Lastly, the oil can reduce oxidative stress in neuronal tissues, preserving cognitive capacity and function.

4. Cardiovascular Disease and Kidney Disease

Molecular or Cellular Target/Pathway: sEH.

Mechanisms: By inhibiting sEH, bay laurel oil can provide significant benefits for cardiovascular and kidney health. By blocking this enzyme, it can increase the levels of protective epoxides that help regulate blood pressure, reduce inflammation, and improve blood flow. This can lead to lower risks of hypertension and atherosclerosis, conditions that can damage the heart and blood vessels. Furthermore, in the kidneys, sEH inhibition may help reduce inflammation and fibrosis, which are common in kidney diseases, ultimately supporting better kidney function and preventing progression to more severe kidney issues. By promoting a healthier cardiovascular and renal system, sEH inhibitors represent a promising approach to enhancing overall heart and kidney health.

5. Antimicrobial (Bacteria, Fungi, Viruses)

Molecular or Cellular Target/Pathway: Bacterial and fungal cell walls/membranes, viral replication.

Mechanisms: Bay laurel disrupts pathogen cell membrane integrity and inhibits microbial growth and replication. Inhibition of viral replication. It has demonstrated broad-spectrum inhibition of a variety of pathogens.

6. Metabolic Regulation (Diabetes, Obesity)

Molecular or Cellular Target/Pathway: Alpha-amylase, alpha-glucosidase, lipase, lipid metabolism pathways.

Mechanisms: Inhibition of carbohydrate-hydrolyzing enzymes and reduction of postprandial glucose spikes by bay laurel oil leads to better glucose management and control. Its inhibition of lipase decreases fat absorption, thereby reducing obesity. Lastly, the oil modulates lipid metabolism in the liver, leading to improved lipid (cholesterol, triglyceride) profiles.

7. Antioxidant

Molecular or Cellular Target/Pathway: ROS, direct antioxidant activity (DPPH, FRAP, ABTS).

Mechanisms: The oil directly scavenges free radicals, reducing oxidative stress. This protects against oxidative damage to cells, tissues, and organs.

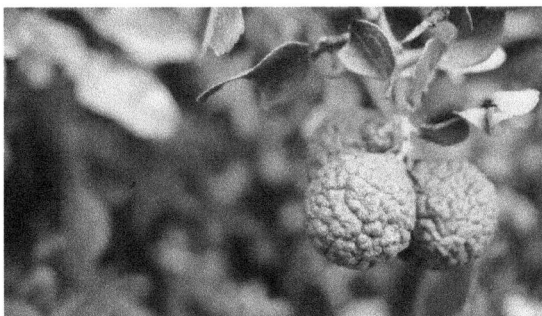

BERGAMOT

Bergamot essential oil, extracted from the distinctive-looking fruit of the bergamot orange (*Citrus bergamia*), has garnered attention for its profound therapeutic potential, attributed to its complex array of bioactive compounds such as limonene, linalyl acetate, and linalool. This essential oil showcases a diverse range of molecular targets and mechanisms of action that position it as a multifaceted natural remedy. Notably, bergamot oil activates autophagy, enhancing cellular housekeeping processes that may contribute to neuroprotection and relief from chronic pain. Additionally, it demonstrates anticancer activity against neuroblastoma and melanoma by inducing apoptosis through mitochondrial pathways and inhibiting cell proliferation. Its cardiovascular benefits stem from vasorelaxation—the process in which blood vessels relax and widen, leading to improved circulation—and regulation of blood pressure, largely mediated by calcium ion modulation and suppression of the hypothalamic-pituitary-adrenal (HPA) axis. Beyond these effects, bergamot oil possesses significant anti-inflammatory and antioxidant capabilities, effectively reducing proinflammatory markers and bolstering the body's natural defense mechanisms. Furthermore, it plays a critical role in neuroprotection and pain relief by maintaining acetylcholine levels and enhancing serotonin activity, while also exhibiting antimicrobial and antiparasitic properties. With its hormonal and mood-regulating effects, including alleviation of stress and postpartum depression, and its application in women's health for dysmenorrhea relief, bergamot oil stands out as a remarkable and versatile natural product with promising therapeutic implications.

SUGGESTED THERAPEUTIC USE:

Conditions Related to Autophagy Activation: Neurodegenerative diseases (Alzheimer's disease, Parkinson's disease, amyotrophic lateral sclerosis, Huntington's disease, multiple sclerosis, frontotemporal dementia, Lewy body dementia, vascular dementia, spinal muscular atrophy, prion diseases), chronic pain, age-related decline

Cancer: Neuroblastoma, melanoma

Conditions Related to Vasorelaxation & Blood Pressure Regulation: Hypertension (High blood pressure), conditions associated with endothelial inflammation (atherosclerosis, diabetes, chronic kidney disease, lupus, rheumatoid arthritis), stress-induced hypertension

Conditions Related to Inflammation and Oxidative Stress: Inflammatory conditions (general), oxidative stress-related conditions (cancer, IBD, diabetes, neurodegenerative conditions, cardiovascular disease, liver or kidney disease, metabolic syndrome, fibromyalgia, chronic fatigue syndrome, skin aging)

Conditions Related to Neuroprotective & Pain-Relieving Effects: Cognitive decline, anxiety, pain management (general, and enhancement of morphine analgesia), neuropathic pain

Bacterial or Parasitic Infections: Bacterial infections (including drug-resistant strains), fungal infections, parasitic infections, insect-borne infections (mosquito larvae, tick infestations)

Hormonal and Mood Balance: Stress, postpartum depression, mood disorders

Conditions Related to Women's Health: Dysmenorrhea (painful menstruation, menstrual cramps), endometriosis, uterine inflammation, oxidative stress related to the reproductive system

MOLECULAR AND CELLULAR TARGETS, BIOLOGICAL PATHWAYS, AND MECHANISMS:

1. Autophagy Activation

Molecular or Cellular Target/Pathway: Enhances cellular housekeeping via autophagy, which may contribute to neuroprotection and chronic pain reduction. Autophagy is the body's cellular recycling system, breaking down and removing damaged components to promote cellular health. This process helps eliminate toxic protein aggregates (e.g., amyloid plaques in Alzheimer's, Lewy bodies in Parkinson's) in the brain, removes damaged mitochondria to enhance energy production, and activates genes linked to longevity and health span.

Mechanism: Induction of autophagy pathways likely involving AMPK activation and mTOR inhibition.

2. Cancer Pathways (Neuroblastoma & Melanoma)

Molecular or Cellular Target/Pathway: Key components (limonene, linalyl acetate, bergamottin, 5-geranyloxy-7-methoxycoumarin) induce apoptosis.

Mechanism: Likely via mitochondrial apoptosis pathways (Bax/Bcl-2, caspase activation) and inhibition of proliferation via cell cycle arrest.

3. Vasorelaxation & Blood Pressure Regulation

Molecular or Cellular Target/Pathway: NO/cGMP pathways, proinflammatory cytokines (TNF-α, IL-6).

Mechanism: Affects calcium ion channels, reducing endothelial inflammation. Calcium ion modulation refers to the regulation of calcium ions in cells. Calcium ions play a crucial role in many cellular processes, including muscle contraction, neurotransmitter release, and cell signaling. Modulation can involve changing the levels of calcium in the cytoplasm or altering how cells respond to calcium, affecting various physiological functions. Lowers stress-induced hypertensive effects by reducing cortisol and corticosterone. Likely targets nitric oxide/cyclic guanosine monophosphate (NO/cGMP) pathways for vasodilation and inhibits proinflammatory cytokines (TNF-α, IL-6). The NO/cGMP pathways play a crucial role in regulating blood pressure. Nitric oxide is produced by endothelial cells lining the blood vessels in response to various stimuli, such as shear stress (the friction from blood flow) and certain signaling molecules. Once produced, NO diffuses into the smooth muscle cells of blood vessels and triggers a series of reactions leading to vasodilation, which means the blood vessels widen. NO stimulates the production of cyclic guanosine monophosphate (cGMP) inside smooth muscle cells. cGMP acts as a second messenger that promotes relaxation and dilation of blood vessels. When blood vessels dilate, resistance to blood flow decreases, leading to a reduction in blood pressure.

4. Anti-Inflammatory & Antioxidant Activity

Molecular or Cellular Target/Pathway: NF-κB and COX-2, PGE2, ROS scavenging, proinflammatory cytokines (IL-1β, IL-6, TNF-α), endogenous antioxidant activity.

Mechanism: Inhibition of NF-κB and COX-2, along with ROS scavenging. Suppresses IL-1β, IL-6, TNF-α, PGE2, and nitric oxide. Boosts endogenous antioxidants (SOD, CAT, GSH) activity.

5. Neuroprotective & Pain-Relieving Effects

Molecular or Cellular Target/Pathway: 5-HT1A receptors, autophagy, NMDA receptor, TRPV1 modulation and desensitization.

Mechanism: Helps maintain acetylcholine levels, beneficial for cognitive function. Enhances 5-HT1A receptor activity, contributing to anxiolytic effects. 5-HT1A receptors are a subtype of serotonin receptors that play significant roles in various physiological and psychological processes. These receptors are found in various areas of the brain, particularly in regions associated with mood regulation, such as the hippocampus, prefrontal cortex, and raphe nuclei. They are also located in peripheral tissues like the heart and GI tract. Activation of 5-HT1A receptors leads to inhibitory effects on neuronal excitability, generally resulting in a reduction of serotonin signaling, which regulates mood (anxiety and depression). Enhances morphine analgesia and reduces pain through autophagy induction. The relationship between autophagy and pain is a complex and emerging area of research. Autophagy can regulate inflammation, which is often a significant contributor to pain. It helps clear damaged cells and reduces proinflammatory cytokines, potentially lowering pain levels associated with inflammatory conditions. Autophagy may also influence the sensitivity of nociceptors—sensory neurons responsible for sensing pain—by regulating the turnover of ion channels and receptors on their membranes. Research suggests that autophagy is involved in neuropathic pain, which results from nerve injury. Dysregulation of autophagy can contribute to neuroinflammation and altered pain perception in these conditions. Likely involves NMDA receptor modulation and TRPV1 desensitization. The NMDA receptor (N-Methyl-D-Aspartate receptor) is a type of glutamate receptor that plays a key role in synaptic plasticity, learning, and memory in the brain. TRPV1 (Transient Receptor Potential Vanilloid 1) is a type of ion channel found in sensory neurons that responds to heat and painful stimuli

6. Broad-Spectrum Antimicrobial & Antiparasitic Effects

Molecular or Cellular Target/Pathway: Drug-resistant *E. faecalis*, *E. faecium*, *Mycoplasma* spp., and *A. flavus*; *H. contortus*; mosquito larvae and *Rh. microplus*.

Mechanism: Disrupts bacterial wall integrity and inhibits aflatoxin production. Limonene potentially acts as an antiparasitic by not only destroying the worm's internal structure but also by damaging its external covering, the tegument. This outer layer is the worm's defense. Bergamot may also inhibit hatching and larval development. *H. contortus* is a blood-sucking parasite of ruminants, such as sheep and goats. It is a major cause of economic loss in the livestock industry, as it can cause anemia, weight loss, and even death in infected animals. Kills mosquito larvae and *Rhipicephalus microplus* ticks.

7. Hormonal & Mood-Regulating Effects

Molecular or Cellular Target/Pathway: GABA receptors and pathways, CRH/HPA axis, ACTH.

Mechanism: Shown to lower cortisol and corticosterone levels, improving resilience to stress. Acts via serotonergic and GABAergic pathways without directly interacting with benzodiazepine receptors.

GABA is the primary inhibitory neurotransmitter in the brain. Its main role is to reduce neuronal excitability throughout the nervous system, which helps to balance the overall activity in the brain. When GABA binds receptors on a neuron, it opens channels that let negatively charged chloride ions enter the neuron or positively charged potassium ions leave, which makes the neuron less likely to send signals. This helps maintain normal brain function and plays a crucial role in mood balance. Likely modulation of CRH/HPA axis, increasing serotonin and GABAergic tone. When you're stressed, your brain releases a hormone called corticotropin-releasing hormone (CRH). CRH travels to a part of your brain called the pituitary gland, which then releases another hormone called ACTH (adrenocorticotropic hormone). ACTH travels through your blood to the adrenal glands, prompting them to release stress hormones, mainly cortisol. This hormone helps your body manage stress by increasing energy and helping with other functions, but too much cortisol over time can be harmful.

8. Women's Health & Dysmenorrhea Relief

Molecular or Cellular Target/Pathway: Oxidative stress, prostaglandins (PGF2α, PGE2), and COX enzymes.

Mechanism: Reduces uterine inflammation, oxidative stress, and prostaglandin (PGF2α, PGE2) overproduction. Prostaglandins (like PGF2α and PGE2) are chemicals in the body that help control various functions, including inflammation and pain. When there is an overproduction of prostaglandins in women's health, it can lead to several issues: menstrual cramps, endometriosis, and chronic inflammation. Likely involves COX inhibition and upregulation of antioxidant defenses.

BLACK PEPPER

Black pepper essential oil, derived from the berries of the *Piper nigrum* plant, is not only a common culinary spice but also a powerhouse of therapeutic benefits. This oil is rich in bioactive compounds like beta-caryophyllene and limonene, which are revered for their diverse range of health-promoting properties. Black pepper oil exhibits potent anti-inflammatory and antioxidant effects by inhibiting proinflammatory cytokines and reducing oxidative stress through the modulation of key molecular pathways. Its analgesic properties are noteworthy, as it can alleviate pain and discomfort by enhancing the effectiveness of conventional analgesics while also exerting neuroprotective effects. Additionally, black pepper essential oil possesses antimicrobial activity, effectively combating a range of bacterial and fungal pathogens. Its stimulating scent not only enhances mood and cognitive function through the modulation of neurotransmitters but also supports respiratory health by acting as a natural expectorant. With its impressive array of pharmacological activities and its ability to support overall well-being, black pepper essential oil stands out as a valuable addition to natural health practices.

SUGGESTED THERAPEUTIC USE:

Physical Performance, Respiratory Health, Cardiovascular Health, and Mood: Improved exercise performance and endurance, enhanced alertness, potential support for cardiovascular health, mood elevation

Nicotine Cravings and Dopamine Regulation: Nicotine addiction and withdrawal support

Pain and Inflammation and Enhanced Thermoregulation: Pain relief (general), inflammatory conditions, thermoregulation support (e.g., managing fevers or chills)

Organ and Tissue Protection: Oxidative stress-related conditions (cancer, IBD, diabetes, neurodegenerative conditions, cardiovascular disease, liver or kidney disease, metabolic syndrome, fibromyalgia, chronic fatigue syndrome, skin aging), cellular protection, brain health support, cardiovascular health support, mitochondrial function support, liver detoxification support, inflammatory control, immune resilience, cancer prevention (potential)

Inflammatory Conditions: Skin inflammation, wound-healing support, conditions involving excessive collagen production (e.g., scarring)

Neurocognitive Deficits and Neuromuscular Function: Cognitive decline, memory enhancement, neuromuscular function support (muscle weakness, fatigue)

DNA Protection & Cancer Prevention: DNA damage prevention, cancer prevention (potential), enhanced detoxification

Gastrointestinal and Gut Microbiome Dysregulation: *H. pylori* infections, gut microbiome modulation, digestive issues

Infections: Bacterial infections, fungal infections

MOLECULAR AND CELLULAR TARGETS, BIOLOGICAL PATHWAYS, AND MECHANISMS:

1. Enhanced Physical Performance, Improved Breathing, Cardiovascular Health, and Mood Enhancement

Molecular or Cellular Target/Pathway: Adrenergic receptor activation.

Mechanism: Enhances alertness and adrenaline release via adrenergic receptor activation. Inhalation increases sympathetic activity and adrenaline levels, indicating activation of beta-adrenergic receptors (sympathetic nervous system). Beta-adrenergic receptors are special proteins found on the surfaces of certain cells in your body. They play a key role in the nervous system, particularly in how your body responds to stress and prepares for action. There are two main types: 1) beta-1 receptors are found in the heart and help increase heart rate and strength of heart contractions; 2) beta-2 receptors are located in other areas of the body like the lungs, where they help relax and open the airways. When activated, beta-adrenergic receptors improve heart rate and strength of heart contractions, which can enhance endurance and exercise performance. The response set in motion by beta-adrenergic activity can contribute to feelings of alertness and well-being, particularly during periods of acute stress or excitement.

2. Reduced Nicotine Cravings and Dopamine Regulation

Molecular or Cellular Target/Pathway: Dopamine pathways, reward circuits, nACh receptors.

Mechanism: Reduction in nicotine cravings suggests interaction with dopamine release and reward circuits. Inhalation decreases nicotine cravings, suggesting modulation of nicotine acetylcholine receptors, possibly through dopamine release regulation. Nicotine acetylcholine receptors (nAChRs) are a type of receptor in the brain and body that primarily respond to the neurotransmitter acetylcholine. Nicotine, the primary addictive substance found in tobacco, interacts with these receptors, leading to various physiological and psychological effects that are central to the development of nicotine addiction. With repeated exposure to nicotine, the brain undergoes neuroadaptive changes. Over time, the number and sensitivity of nAChRs can increase or decrease as the brain tries to maintain homeostasis. This leads to tolerance, where the individual requires larger doses of nicotine to achieve the same pleasurable effects. Nicotine acetylcholine receptors play a crucial role in the development and maintenance of nicotine addiction. Their activation leads to increased dopamine release, contributing to the pleasurable sensations that encourage repeated use.

3. Diminished Pain and Inflammation, and Enhance Thermoregulation

Molecular or Cellular Target/Pathway: CB2 receptors, modulation of TRPV1, proinflammatory cytokines (TNF-α, IL-6, IL-1β), COX-2.

Mechanism: Contains beta-caryophyllene, an agonist of CB2 receptors and a modulator of TRPV1 (vanilloid receptors), influencing pain, inflammation, and thermoregulation. Beta-caryophyllene selectively activates CB2 receptors, influencing pain perception, inflammation, and immune response. Suppression of proinflammatory cytokines (TNF-α, IL-6, IL-1β) and inhibition of COX-2 and PGE2 synthesis contribute to pain relief. The mechanism of action of beta-caryophyllene involves multiple pathways, including its interactions with the cannabinoid system and thermoregulation. Selective binding to and activation of CB2 receptors, which inhibits the release of proinflammatory cytokines. Modulation of TRPV1 receptors, influencing pain sensation and thermoregulation.

4. Organ and Tissue Protection

Molecular or Cellular Target/Pathway: Endogenous antioxidants (SOD, GSH, CAT, GSH-R).

Mechanisms: Upregulation of superoxide dismutase (SOD), glutathione (GSH), catalase (CAT), and glutathione reductase (GSH-R) suggests involvement in redox balance and oxidative stress reduction. This enhances the body's antioxidant defense system. Leading to protection of cells, the brain, cardiovascular health, mitochondrial function, and improved liver detoxification and toxin defense, inflammatory control, immune resilience, and cancer prevention.

5. Anti-Inflammatory Effects

Molecular or Cellular Target/Pathway: NF-κB, COX-2, and PPAR-γ signaling pathways and dermal fibroblasts.

Mechanism: Inhibition of cytokine-stimulated dermal fibroblasts and collagen production suggests modulation of NF-κB, COX-2, and PPAR-γ signaling pathways, contributing to anti-inflammatory effects. Cytokine-stimulated dermal fibroblasts play a crucial role in skin health and wound healing, primarily through their involvement in collagen production and the inflammatory response. Dermal fibroblasts are specialized cells located in the dermis (the inner layer of the skin), primarily responsible for producing collagen, elastin, and other extracellular matrix components, which give the skin its strength, elasticity, and overall structure. Targeting this pathway could help modulate inflammation and promote healthy collagen deposition, improving skin health and recovery processes.

6. Brain and Cognitive Benefits, and Improved Neuromuscular Function

Molecular or Cellular Target/Pathway: AChE receptors and pathways.

Mechanism: Black pepper oil inhibits AChE, increasing acetylcholine availability, which could enhance cognitive function and memory and neuromuscular function. Acetylcholine is responsible for activating muscles, ensuring smooth movement and coordination and preventing muscle weakness and fatigue by maintaining nerve-to-muscle communication.

7. DNA Protection & Cancer Prevention

Molecular or Cellular Target/Pathway: Detoxification enzymes (GST, UGT), Nrf2/ARE signaling, telomere length.

Mechanism: Inhibition of DNA adduct formation suggests involvement in phase II detoxification enzymes (e.g., glutathione-S-transferase, UDP-glucuronosyl-transferase). DNA adducts are formed when reactive chemicals bind covalently to DNA, altering its structure and potentially disrupting normal cellular functions. DNA adducts can lead to mutations during DNA replication, which may disrupt normal cellular functions. Mutations in critical genes (such as tumor suppressor genes or oncogenes) can initiate cancer development. The presence of DNA adducts can lead to genetic instability, which is a hallmark of many cancers. DNA damage may trigger inflammatory responses, contributing to a cycle of inflammation and additional DNA damage. Upregulation of Nrf2/ARE (Antioxidant Response Element) signaling, leading to increased production of detoxification and antioxidant enzymes. Black pepper oil increases telomere length, possibly through telomerase activation and oxidative stress reduction.

8. Gastrointestinal and Gut Microbiome Effects

Molecular or Cellular Target/Pathway: *H. pylori*.

Mechanism: Moderate anti-*Helicobacter pylori* activity, potential gut microbiome modulation, and digestive enzyme activation. *H. pylori* is a gram-negative bacterium that primarily infects the stomach lining. It is known to be a significant contributor to various gastrointestinal diseases including peptic ulcers, gastritis, dyspepsia, and gastric cancer. Research has also linked *H. pylori* infection to iron deficiency anemia, cardiovascular disease, and dysregulated metabolic function.

9. Antimicrobial & Antifungal

Molecular or Cellular Target/Pathway: Pathogen membrane disruption, quorum sensing, efflux pump.

Mechanism: Effective against various bacteria and fungi, likely through membrane disruption, quorum sensing inhibition, and efflux pump inhibition.

BLUE TANSY

With a deep blue hue due to its chamazulene content, blue tansy (*Tanacetum annum*) exhibits notable anticancer, anti-inflammatory, and antifungal activities. Widely used in aromatherapy and skincare, blue tansy is valued for its calming effects on the mind and body. Its soothing aroma promotes relaxation, reduces stress, and supports emotional balance. In skin care, blue tansy helps calm irritated skin, making it beneficial for conditions like eczema, acne, and redness. Its antioxidant properties also protect against environmental stressors. Additionally, blue tansy has antimicrobial and anti-inflammatory effects, which may support overall health. Whether used in a diffuser, blended into a carrier oil, or added to skincare formulations, blue tansy essential oil is a versatile natural remedy for both emotional and physical well-being.

SUGGESTED THERAPEUTIC USE:

Inflammatory Conditions: Chronic inflammatory conditions (rheumatoid arthritis, IBD, psoriasis, COPD, asthma, lupus, eczema, gout, chronic sinusitis, fibromyalgia, chronic fatigue syndrome, periodontitis), conditions with excessive leukotriene production (asthma, seasonal allergies, COPD, eosinophilic esophagitis, psoriasis, allergic conjunctivitis, lupus, cystic fibrosis, gout), allergic reactions

Skin Conditions: Eczema, psoriasis, irritated skin, wounds

Cancer: Rhabdomyosarcoma

Infections: Fungal

MOLECULAR AND CELLULAR TARGETS, BIOLOGICAL PATHWAYS, AND MECHANISMS:

1. **Anticancer Activity**

Molecular or Cellular Target/Pathway: Direct cytotoxicity to cancer cells.

Mechanism: Blue tansy oil demonstrated cytotoxic activity against a human rhabdomyosarcoma cancerous cell line. Monoterpenes may inhibit cancer cell proliferation by preventing protein isoprenylation, inhibiting DNA biosynthesis, or inducing apoptosis. Protein isoprenylation involves attaching a specific "tag" to a protein that helps it do its job in the cell. In cancer cells, certain proteins rely heavily on isoprenylation to function. These proteins are often involved in processes that allow cancer cells to grow uncontrollably, invade other tissues, and avoid destruction. By blocking isoprenylation, anticancer treatments can disrupt these crucial functions, making it harder for cancer cells to survive and spread. Cancer cells divide much more rapidly than healthy cells. DNA biosynthesis is essential for this rapid division because it provides the necessary DNA for the new cells. By inhibiting DNA biosynthesis, anticancer agents can slow down or stop cancer cell division. This is particularly effective because cancer cells are much more sensitive to disruptions in DNA replication than normal cells.

2. Anti-inflammatory

Molecular or Cellular Target/Pathway: 5-LOX.

Mechanism: Blue tansy mildly inhibits the 5-LOX enzyme, reducing inflammation.

3. Antifungal Activity

Molecular or Cellular Target/Pathway: Fungal inhibition.

Mechanism: Blue tansy oil inhibits the growth of various fungi, especially those that grow in foods.

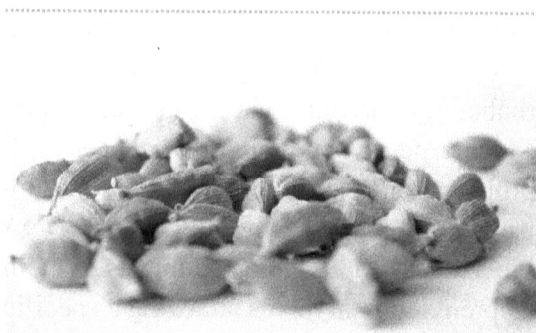

CARDAMOM

Cardamom essential oil, extracted from the aromatic seeds of *Elettaria cardamomum*, is a treasure trove of therapeutic properties that offers a wide array of health benefits. Renowned for its potential anticancer mechanisms, this oil demonstrates chemopreventive effects by inhibiting DNA adduct formation, thus preventing the binding of carcinogenic chemicals to DNA. Its impressive antioxidant activity combats oxidative stress, a core contributor to cancer development, by targeting key molecular pathways such as NF-κB and p53. In addition to its anticancer properties, cardamom essential oil serves as a gastroprotective agent, showing superior efficacy in preventing stomach ulcers compared to conventional medication like ranitidine, likely through the inhibition of gastric acid secretion and enhancement of mucosal defense. The oil also exerts systemic anti-inflammatory and immune-modulating effects by suppressing proinflammatory cytokines, further solidifying its role in promoting overall health. Additionally, its broad-spectrum antimicrobial and antiparasitic activities, combined with neuromodulatory effects that may bolster cognitive function, position cardamom essential oil as a versatile and potent ally in both traditional and modern health practices. With its remarkable ability to alleviate gastrointestinal issues, reduce nausea, and repel pests, cardamom essential oil is an essential component of natural wellness through its multifaceted therapeutic actions.

SUGGESTED THERAPEUTIC USE:

Cancer: Cancer prevention (chemopreventive effects), reduction of oxidative stress related to cancer, support in regulating cell proliferation and apoptosis

Gastrointestinal Protection: Stomach ulcers, gastric mucosal inflammation, excessive gastric acid secretion

Inflammatory Conditions and Dysregulated Immune Function: General inflammation, conditions involving excessive proinflammatory cytokines (rheumatoid arthritis, IBD, psoriasis, asthma, COPD, diabetes, obesity, multiple sclerosis, Alzheimer's disease, fibromyalgia, gout), immune system modulation

Microbial and Parasitic Infections: Bacterial infections, protozoal infections, parasitic infections

Neurocognitive Conditions: Cognitive decline, memory enhancement, neurodegeneration-related conditions (Alzheimer's disease, Parkinson's disease, amyotrophic lateral sclerosis, Huntington's disease, multiple sclerosis, frontotemporal dementia, Lewy body dementia, vascular dementia, spinal muscular atrophy, prion diseases)

Gastrointestinal Regulation: Diarrhea, intestinal spasms

Antiemetic: Nausea, vomiting

Insecticidal and Repellent Effects: Insect infestations, tick infestations, mosquito larvae control

MOLECULAR AND CELLULAR TARGETS, BIOLOGICAL PATHWAYS, AND MECHANISMS:

1. Anticancer

Molecular or Cellular Target/Pathway: DNA adduct formation, antioxidant activity, NF-κB and p53.

Mechanism: Cardamom essential oil may inhibit DNA adduct formation, preventing cancer-causing chemicals from binding to DNA (suggesting a chemopreventive effect). Demonstrated in DPPH and TEAC assays, which may reduce oxidative stress,

a key driver of carcinogenesis. NF-κB and p53 are two important proteins that play crucial roles in regulating cell proliferation (the process of cells dividing and growing) and apoptosis (programmed cell death). When activated, NF-κB can promote cell proliferation by inducing the expression of genes that are involved in cell growth and survival. This means it can counteract programmed cell death, allowing cells to live longer under stress or inflammatory conditions. p53 is often referred to as the "guardian of the genome" because it helps maintain genomic integrity by responding to cellular stress or damage. p53 can halt the cell cycle (the process of cell division) when DNA damage is detected, giving the cell time to repair the damage. If the damage is irreparable, it will trigger cell death.

2. Gastroprotective

Molecular or Cellular Target/Pathway: H+/K+ ATPase and COX-2, mucosal defense, and gastric acid secretion.

Mechanisms: Cardamom extracts showed better protection against stomach ulcers than Zantac in animal models. Likely involves inhibition of gastric acid secretion and enhancement of mucosal defense. H+/K+ ATPase, often referred to as the proton pump, is an important protein found in the stomach lining. It helps control the acidity in our stomachs, which is crucial for digestion and protecting our stomachs from harmful substances. It is located on the surface of cells in the stomach lining. When you eat, the proton pump goes to work, pushing hydrogen ions into the stomach. This creates an acidic environment, which is essential for breaking down food and killing harmful bacteria. The pump also helps balance the amount of acid in the stomach. Too much acid can be harmful, so the body regulates the activity of the proton pump to ensure levels are just right. By balancing acid levels and contributing to the protective mucus layer, it plays a key role in keeping our stomach healthy and functioning well.

3. Anti-inflammatory and Immune Modulation

Molecular or Cellular Target/Pathway: NF-κB and COX-2, both of which are central to inflammation pathways.

Mechanisms: Decreased proinflammatory molecule secretion beyond the gastrointestinal tract in animal models. Likely through suppression of proinflammatory cytokines such as TNF-α, IL-6, and IL-1β.

4. Broad-Spectrum Antimicrobial and Antiparasitic Activity

Molecular or Cellular Target/Pathway: Pathogen cell wall/membrane disruption, inhibition of biofilm formation, and immune stimulation; parasitic inhibition; and larvicidal.

Mechanism: Active against *Toxoplasma gondii*, likely through immune modulation and oxidative stress reduction. A single-celled parasitic organism, *T. gondii* can infect many warm-blooded animals, including humans (called toxoplasmosis). It is best known for its ability to live in cats, which are the primary hosts for this parasite. People can become infected with *T. gondii* through various ways, such as consuming undercooked meat, handling cat litter, or coming into contact with contaminated soil or water. Most healthy individuals experience mild or no symptoms, but the parasite can cause serious health issues for pregnant women or those with weakened immune systems, as it can affect the fetus or lead to complications in such individuals. Larvicidal effects against *An. stephensi* mosquito larvae. Inhibits *Echinococcus granulosus* (hydatid worm), with 1,8-cineole enhancing its effect when combined with albendazole. The hydatid worm can infect humans, leading to a condition known as echinococcosis or hydatid disease. Once ingested, the eggs hatch in the intestines and release larvae that migrate to various organs, most commonly the liver and lungs, where they can form cysts. These cysts can grow slowly over time and may cause symptoms depending on their size and location, including abdominal pain, nausea, or allergic reactions. In some cases, rupturing of the cysts can lead to serious complications.

5. Neuromodulatory Effects

Molecular or Cellular Target/Pathway: AChE and BChE

Mechanism: Inhibits AChE and BChE due to alpha-terpinyl acetate, 1,8-cineole, linalool, linalyl acetate, and alpha-terpineol, suggesting potential cognitive and neuroprotective benefits. May support memory and neurodegeneration-related conditions such as Alzheimer's disease.

6. Gastrointestinal Regulation

Molecular or Cellular Target/Pathway: Modulation of calcium channels and inhibition of acetylcholine-induced contractions.

Mechanisms: Cardamom oil reduced castor oil-induced diarrhea and intestinal spasms, possibly through modulation of calcium channels and inhibition of acetylcholine-induced contractions. Calcium channels play a crucial role in the contraction of smooth muscle in the gastrointestinal tract. When acetylcholine binds to its receptors on smooth muscle cells, it triggers the opening of calcium channels, allowing calcium ions to enter the cells, which leads to muscle contraction and, consequently, increased motility. In cases of diarrhea, excessive stimulation of these pathways can lead to overactive intestinal contractions and cramping. Inhibition of acetylcholine-induced contractions can help manage symptoms by reducing calcium influx and muscle contraction, thereby slowing down gut motility and alleviating diarrhea and associated cramping.

7. Antiemetic Activity

Molecular or Cellular Target/Pathway: 5-HT3 receptors and vagal pathways.

Mechanisms: Inhalation of cardamom essential oil during cesarean sections significantly lowered nausea severity. Likely interacts with serotonin (5-HT3) receptors and vagal pathways regulating nausea. 5HT-3 receptors are found in the central nervous system and the gastrointestinal (GI) tract. When the intestines are irritated or when there is a presence of certain emetic substances, serotonin is released from enterochromaffin cells in the gut. This serotonin can then activate 5-HT3 receptors on vagal afferent neurons, which transmit signals to the brain, specifically to the area postrema and the nucleus tractus solitarius, leading to the sensation of nausea and triggering the vomiting reflex. The vagus nerve carries afferent signals from the gut to the central nervous system. The brain interprets these signals as nausea, and if the signals are strong enough, it can initiate the vomiting response to rid the body of potential toxins or irritants. Together, the interaction between serotonin signaling and vagal pathways is crucial for the body's ability to detect and respond to gastrointestinal disturbances that may lead to nausea and vomiting.

8. Insecticidal and Repellent Effects

Molecular or Cellular Target/Pathway: Neuromuscular and respiratory functions of parasites/ticks.

Mechanisms: Disruption of insect neurotransmission and metabolic pathways.

CASSIA

Cassia essential oil, derived from the bark of *Cinnamomum cassia*, presents a broad spectrum of therapeutic potential. Its potent antimicrobial activity, particularly against drug-resistant pathogens like *Staphylococcus aureus* and *Klebsiella pneumoniae*, underscores its clinical relevance. Furthermore, its efficacy in reducing inflammation and modulating immune responses suggests a role in managing inflammatory conditions. Cassia's ability to lower blood glucose and improve insulin sensitivity highlights its potential in diabetes management. Additionally, its observed antiviral and anticancer activities, coupled with its ability to enhance antibiotic effectiveness, indicate a complex interplay with cellular pathways. The variability in its constituent composition across seasons, as observed in leaf oil, further emphasizes the need for standardized extraction and analysis to optimize its therapeutic applications.

SUGGESTED THERAPEUTIC USE:

Cancer: Breast, Lung

Infections: Bacterial, antibiotic-resistant infections, fungal, viral, parasitic, *C. diff*, respiratory tract infections

Metabolic Disorders: Type 2 diabetes, insulin resistance, metabolic disorder

Inflammatory Conditions: General inflammatory conditions, chronic inflammatory conditions (rheumatoid arthritis, IBD, psoriasis, COPD, asthma, lupus, eczema, gout, chronic sinusitis, fibromyalgia, chronic fatigue syndrome, periodontitis)

Menstrual Pain: Menstrual cramps and pain

Skin Conditions: Acne, wounds, skin brightening, hyperpigmentation, skin infections

Pain Management: General pain management

Male Reproductive Issues: Erectile dysfunction

Cognitive or Neurological Disorders: Protection against neurotoxicity, Parkinson's disease

MOLECULAR AND CELLULAR TARGETS, BIOLOGICAL PATHWAYS, AND MECHANISMS:

1. **Anti-inflammatory Effects**

Molecular or Cellular Target/Pathway: iNOS, COX-2, PGE2, TNF-α, proinflammatory cytokines and chemokines (IL-1β, IL-6, IL-8), NF-κB, activation of anti-inflammatory mediators (IL-10, TGF-β), JAK/STAT signaling pathways.

Mechanisms: Cinnamaldehyde, a major constituent of cassia oil, suppresses the synthesis and release of proinflammatory cytokines and enzymes. It simultaneously increases the activation of anti-inflammatory mediators, like IL-10 and, transforming growth factor-beta (TGF-β). When TGF-β is activated, it signals your immune system to calm down. It tells the cells causing inflammation to stop producing so many inflammatory chemicals and also encourages the creation of cells that help repair damaged tissue and prevent further immune overreactions. Essentially, TGF-β helps restore balance, shifting the body from a state of inflammatory chaos back to a state of calm and healing. Cassia oil modulates signaling pathways involved in inflammation, such as the NF-κB and JAK/STAT pathways, thereby reducing inflammatory responses and joint swelling. Cassia oil can also improve macrophage polarization. Macrophages are like the cleanup crew of your immune system, and they can have different types and perform different roles. Some promote inflammation, while others help resolve it. Improving macrophage polarization means shifting these cells from their proinflammatory state to an anti-inflammatory one. So, instead of throwing fuel on the fire, they start to dampen it down, helping to clear debris and promote healing. Lastly, the oil, through its various mechanisms (inhibiting proinflammatory cytokines, modulating signaling pathways, and affecting macrophage polarization), was effective in a model of rheumatoid arthritis. It decreased the production of inflammatory mediators, lessened the recruitment of immune cells to the joint tissues, and reduced the fluid accumulation and tissue damage that caused the swelling.

2. Antimicrobial and Antiviral Activity

Molecular or Cellular Target/Pathway: Pathogen cell walls/ membranes, biofilm formation.

Mechanisms: Cassia oil, particularly trans-cinnamaldehyde, disrupts the integrity of bacterial and fungal cell membranes, leading to cell lysis. It inhibits biofilm formation, reducing bacterial resistance and persistence. The oil also inhibits the growth of viruses (influenza A virus).

3. Antidiabetic Effects

Molecular or Cellular Target/Pathway: Pancreatic beta-cells (INS-1 cells), insulin-stimulated tyrosine phosphorylation (IR-β, IRS-1), blood glucose levels.

Mechanisms: Cassia oil stimulates insulin secretion from pancreatic beta-cells (INS-1 cells). Cassia bark oil helps the body manage blood sugar by boosting the function of insulin, the hormone that moves sugar from the blood into cells for energy. It does this in two key ways. First, it stimulates special cells in the pancreas (called beta cells) to release more insulin, ensuring there's enough of it available when needed. Second, it makes the body's cells more responsive to insulin by improving a process called tyrosine phosphorylation, which acts like a signal switch, allowing insulin to work more effectively. This means sugar is cleared from the blood more efficiently, preventing high blood sugar spikes and reducing the risk of diabetes-related problems. Over time, this can lead to better energy levels, balanced metabolism, and overall improved health.

4. Anticancer Activity

Molecular or Cellular Target/Pathway: Cytotoxic to breast and lung cancer cells.

Mechanisms: Cassia essential oil-impregnated chitosan nanoparticles enhance antitumor activity and reduce cancer cell viability.

5. Neuroprotective Effects

Molecular or Cellular Target/Pathway: Dopamine-producing neurons, oxidative stress reduction.

Mechanisms: Cassia oil protects neurons against toxicity and cell death, potentially by reducing oxidative stress and preserving neuronal function. The model it was used in showed it protected against 6-hydroxydopamine—a neurotoxic chemical that researchers use to study Parkinson's disease—toxicity.

6. Skin Pigmentation Effects

Molecular or Cellular Target/Pathway: Tyrosinase.

Mechanism: Cassia oil inhibits tyrosinase, reducing melanin production and therefore brightening the skin and reducing hyperpigmentation.

CEDARWOOD (ATLAS/HIMALAYAN)

Cedarwood essential oil, derived from both Atlas (*Cedrus atlantica*) and Himalayan (*C. deodora*) cedar trees, exhibits a diverse range of therapeutic properties. They are both rich in himachalenes (beta, alpha, and gamma) with moderate levels of atlantones (gamma and alpha) and himachalols (alpha, beta, and allo). Atlas cedarwood tends to have higher levels of beta-himachalene, while Himalayan cedarwood normally is higher in atlantones and himachalols. Research highlights their potential in modulating inflammation, pain, and immune responses alongside promising applications in cancer and dermatological care. Studies indicate that cedarwood can inhibit myeloid leukemia cell spread and modulate inflammatory pathways by targeting the 5-LOX enzyme. Furthermore, it demonstrates efficacy in reducing stomach acidity and protecting the gastric lining. Notably, Himalayan cedarwood shows enhanced anti-inflammatory and analgesic effects compared to its Atlas counterpart. The oil's capacity to balance mast cell activity and inhibit leukotriene production suggests potential benefits in managing allergic and autoimmune disorders. Additionally, clinical and preclinical studies document its effectiveness in pain relief, wound healing, and even tick eradication, positioning cedarwood as a versatile natural remedy.

SUGGESTED THERAPEUTIC USE:

Cancer: Colon, myeloid leukemia

Inflammatory Conditions: Gastrointestinal inflammation, chronic inflammatory conditions (rheumatoid arthritis, IBD, psoriasis, COPD, asthma, lupus, eczema, gout, chronic sinusitis, fibromyalgia, chronic fatigue syndrome, periodontitis), conditions with excessive leukotriene production (asthma, seasonal allergies, COPD, eosinophilic esophagitis, psoriasis, allergic conjunctivitis, lupus, cystic fibrosis, gout), general inflammation

Skin Conditions: Eczema, dermatitis, hair loss (alopecia), acne, skin tone and pigmentation

Pain: General pain relief, postoperative pain, muscle aches and pains

Immune Disorders: Allergic conditions, asthma, autoimmune disorders, multiple sclerosis

Gastrointestinal Disorders: GERD, heartburn, GI inflammation

Infections: Bacterial, fungal

Pest Infections: Ticks

MOLECULAR AND CELLULAR TARGETS, BIOLOGICAL PATHWAYS, AND MECHANISMS:

1. Anti-inflammatory Effects

Molecular or Cellular Target/Pathway: 5-LOX, mast cells, leukotriene production, NF-κB.

Mechanisms: Cedarwood essential oil inhibits the 5-LOX enzyme and reduces leukotriene synthesis, thereby decreasing inflammation. Mast cell stabilization prevents the release of inflammatory mediators like histamine. Inhibition of NF-κB reduces the transcription of proinflammatory genes.

2. Analgesic (Pain-Relieving) Effects

Molecular or Cellular Target/Pathway: Opioidergic, serotonergic, noradrenergic, and dopaminergic systems, and the endocannabinoid system.

Mechanisms: Cedarwood exerts pain-relieving effects through multiple mechanisms and biological targets. It influences opioidergic, serotonergic, noradrenergic, and dopaminergic systems, and the endocannabinoid system. All of these are key pathways in pain signaling. Cedarwood essentially alters how pain signals are processed in the brain and spinal cord. Activation of descending pain modulation pathways through neurotransmitter systems. Modulation of the endocannabinoid system, potentially by inhibiting enzymes that degrade endocannabinoids, enhances pain relief.

3. Anticancer Effects

Molecular or Cellular Target/Pathway: NF-κB, colon cancer cell proliferation.

Mechanisms: The inhibition of NF-κB by cedarwood suppresses cancer cell proliferation, apoptosis resistance, angiogenesis, and metastasis.

4. Immunomodulatory Effects

Molecular or Cellular Target/Pathway: Humoral and cell-mediated immune responses.

Mechanisms: Through its modulation of immune cell activity, cedarwood potentially influences cytokine production and immune cell signaling. The oil changes how immune cells talk to each other using chemical messengers called cytokines. If the immune system is overreacting (like in allergies or autoimmune diseases), it helps calm these responses. If the immune system is underreacting, the oil can boost activity.

5. Antimicrobial and Antifungal Effects

Molecular or Cellular Target/Pathway: Bacterial and fungal cell walls/membranes.

Mechanisms: Cedarwood disrupts microbial cell membranes, inhibiting growth. Inhibition of tyrosinase reduces melanin production, leading to skin brightening.

6. Skin Pigmentation Effects

Molecular or Cellular Target/Pathway: Tyrosinase.

Mechanisms: By inhibiting tyrosinase, an enzyme involved in melanin production, cedarwood oil reduces the synthesis of dark pigments, leading to a more even and brighter skin tone.

CEDARWOOD (VIRGINIA)

Virginia cedarwood (*Juniperus virginiana*) essential oil, extracted from the wood of the tree, is highly esteemed for its diverse array of therapeutic properties that support both physical and emotional health. Its remarkable wound-healing capabilities are illustrated through its potential to enhance cellular proliferation and migration, likely promoting collagen synthesis and speeding up epithelialization processes, as demonstrated in studies involving rat models. The oil also showcases effective insect-repellent properties, with its key constituents, such as cedrol and thujopsene, disrupting insect olfactory sensory pathways and inducing avoidance behaviors against pests like fire ants. Additionally, Virginia cedarwood essential oil exhibits potent anti-inflammatory effects by targeting the 5-LOX enzyme, thereby inhibiting proinflammatory leukotrienes, which can alleviate chronic inflammation. Beyond these benefits, its anxiety-relieving properties are tied to modulating key neurotransmitters such as serotonin and dopamine, offering soothing support for mood regulation. Furthermore, the oil's antibacterial action against *H. pylori* and its antimicrobial effects against various pathogens underscore its versatility as a natural remedy. With larvicidal properties and contact toxicity against insect larvae, Virginia cedarwood essential oil emerges as a multifaceted agent that not only promotes healing but also provides protective benefits across various health domains.

SUGGESTED THERAPEUTIC USE:

Wounds and Abrasions: Minor wounds, skin abrasions, promoting collagen synthesis, epithelialization support

(wounds, burns, skin grafts, oral ulcers, gastrointestinal tract mucosal lesions, surgical incisions)

Insect Repellence: Repelling fire ants, repelling red bud borers, general insect repellent

Inflammatory Conditions: Chronic inflammatory conditions (rheumatoid arthritis, IBD, psoriasis, COPD, asthma, lupus, eczema, gout, chronic sinusitis, fibromyalgia, chronic fatigue syndrome, periodontitis), conditions with excessive leukotriene production (asthma, seasonal allergies, COPD, eosinophilic esophagitis, psoriasis, allergic conjunctivitis, lupus, cystic fibrosis, gout), general inflammation

Mood and Stress: Anxiety, stress, mood regulation

***H. pylori* Infections**: *H. pylori* infections, potential support in reducing gastritis and ulcers, potential support in reducing the risk of *H. pylori*-related gastric cancers

Insect Control: Insect larvae control, pest control

Infections: Bacterial infections, fungal infections, general antimicrobial support

MOLECULAR AND CELLULAR TARGETS, BIOLOGICAL PATHWAYS, AND MECHANISMS:

1. Wound-Healing Mechanism

Molecular or Cellular Target/Pathway: The specific molecular targets in the context of wound healing are not explicitly detailed, but it can be inferred that growth factors and cytokines play a role.

Mechanism: The application of Virginia Cedarwood essential oil may enhance pathways related to cellular proliferation and migration (e.g., Wnt/β-catenin signaling, TGF-β signaling). The oil likely promotes collagen synthesis and epithelialization, contributing to improved wound healing as evidenced by reduced healing times in the studied model (rats).

2. Insect Repellent Properties

Molecular or Cellular Target/Pathway: Virginia Cedarwood oil's constituents (high in cedrol and thujopsene) may interact with olfactory receptors in insects.

Mechanism: The oil's repellent action against pests such as fire ants and the redbud borer is likely due to its ability to disrupt sensory perception or interfere with their nervous systems.

3. Anti-Inflammatory Effects

Molecular or Cellular Target/Pathway: 5-LOX enzyme is the primary target.

Mechanism: By inhibiting the 5-LOX pathway, Virginia Cedarwood oil reduces the conversion of arachidonic acid into leukotrienes, which are proinflammatory mediators. This inhibition contributes to decreased inflammation, which can have implications for chronic inflammatory conditions and conditions exacerbated by excessive leukotriene production.

4. Anxiolytic Effects

Molecular or Cellular Target/Pathway: Key neurotransmitters such as serotonin (5-HT) and dopamine are major targets.

Mechanism: The anxiolytic effect may arise from altered neurotransmission in brain areas involved in mood regulation, such as the hippocampus and hypothalamus. Virginia Cedarwood oil and its constituent cedrol reportedly increase serotonin levels while reducing dopamine, potentially influencing serotonergic and dopaminergic pathways in the brain. Serotonin and dopamine are neurotransmitters that significantly influence mood and anxiety regulation. Serotonin is often associated with feelings of well-being and stability; low levels of serotonin have been linked to increased anxiety and mood disorders. Dopamine, on the other hand, is involved in reward processing and motivation, and dysregulation of dopamine pathways can contribute to anxiety symptoms and exacerbate stress responses. Together, the balance between these neurotransmitters plays a critical role in emotional regulation and the experience of anxiety.

5. Antibacterial Activity Against *Helicobacter Pylori*

Molecular or Cellular Target/Pathway: The urease enzyme of *H. pylori* is particularly targeted.

Mechanism: Urease is a crucial enzyme produced by the bacterium that impacts the bacterium's survival and pathogenicity in the acidic environment of the stomach. Inhibition of urease disrupts *H. pylori's* ability to survive in the acidic environment of the stomach, preventing the development of gastritis and ulcers. This makes Virginia Cedarwood essential oil effective in reducing the viability of H. pylori and potentially lowering the risk of related gastric cancers.

6. Larvicidal and Contact Toxicity

Molecular or Cellular Target/Pathway: The signaling pathways and cellular structures in insect larvae.

Pathways: Disruption of neural signaling pathways that lead to paralysis or lethargy in target insects. By disrupting the physiological processes of larvae, mortality through contact toxicity occurs.

7. Antimicrobial Activity

Molecular or Cellular Target/Pathway: Cell walls and membranes of various bacteria and fungi.

Mechanism: Compounds in Virginia Cedarwood essential oil may compromise cellular integrity or inhibit metabolic pathways in microbes. While the potency is described as weak, the oil can still exert antimicrobial effects through direct action on target pathogens.

CINNAMON BARK/LEAF

Cinnamon essential oil, derived from the bark or leaf of *Cinnamomum verum* (syn. *C. zeylanicum*), is a potent oil celebrated for its extensive antimicrobial, anti-inflammatory, and antioxidant properties. Its remarkable antimicrobial activity targets bacterial cell membranes, leading to cellular leakage and disrupting biofilm formation among notorious pathogens such as *S. aureus* and *E. coli*. The oil's antifungal potential is equally impressive; it inhibits ergosterol biosynthesis in fungi and alters mycelial growth pathways, effectively combating fungal infections. Beyond its role in pathogen defense, cinnamon essential oil exhibits anti-inflammatory effects by modulating proinflammatory cytokines and inhibiting the NF-κB signaling pathway, while its antioxidant properties protect tissues from oxidative damage by scavenging reactive oxygen species. Moreover, this essential oil has been shown to enhance cognitive function by inhibiting acetylcholinesterase, contributing to neuroprotective benefits, and displays potential antidiabetic effects by improving glucose metabolism and insulin sensitivity. With its multifaceted mechanisms of action ranging from wound healing and moisture retention to anticancer and respiratory protective effects, cinnamon essential oil serves as a remarkable natural remedy, promising a wealth of health benefits for both topical and internal applications.

SUGGESTED THERAPEUTIC USE:

Bacterial Infections: Bacterial infections, biofilm-related infections (chronic wound infections, catheter-associated UTIs, heart valve infections, chronic sinusitis, cystic fibrosis, dental plaque, periodontal disease, osteomyelitis, conjunctivitis, implant-associated infections, ventilator-associated pneumonia, dialysis-associated infections), multidrug-resistant bacterial infections

Fungal Infections: Fungal infections, ergosterol biosynthesis-related fungal issues

Inflammatory Conditions: Inflammatory conditions, cytokine-related disorders

Organ and Tissue Protection: Oxidative stress-related conditions (cancer, IBD, diabetes, neurodegenerative conditions, cardiovascular disease, liver or kidney disease, metabolic syndrome, fibromyalgia, chronic fatigue syndrome, skin aging), free radical damage

Metabolic Disorders: Diabetes, hyperglycemia, impaired glucose tolerance, compromised insulin sensitivity

Cognitive and Neurological Decline: Cognitive decline, memory impairment, neurodegenerative diseases

Cancer: Certain cancers (e.g., prostate, breast)

Respiratory Conditions: Influenza, viral respiratory infections

Wounds: Wounds, skin injuries, impaired wound healing

MOLECULAR AND CELLULAR TARGETS, BIOLOGICAL PATHWAYS, AND MECHANISMS:

1. Broad-Spectrum Antibacterial Activity

Molecular or Cellular Target/Pathway: Bacterial cell membrane damage, leading to leakage of intracellular contents;

disruption of pathways involved in biofilm formation; suppression of toxin production.

Mechanism: Cinnamon inhibits quorum sensing, a bacterial communication system that regulates biofilm formation and virulence. Bacterial quorum sensing is a way for bacteria to communicate with each other using small signaling molecules. When bacteria are in a low population, they produce these signals at a low level, so they don't affect much. But as they grow and the number of bacteria increases, they begin to release more of these molecules. Once the concentration of signals reaches a certain level, the bacteria detect them and "realize" there are enough of them nearby. This leads them to change their behavior as a group, like forming biofilms, producing toxins, or starting to move together. Essentially, quorum sensing allows bacteria to work together more effectively, like teamwork for survival and adaptation. Cinnamon has been shown to exert strong antimicrobial effects against a spectrum of bacteria and fungi by disrupting their cell membranes, inhibiting virulence factors (e.g., Shiga toxin production), and preventing biofilm formation and maturation. Bacterial virulence factors are special tools that help bacteria infect the body and make people sick. These include toxins (poisons that damage cells), capsules (slimy coatings that protect bacteria from the immune system), adhesins (sticky proteins that help bacteria attach to surfaces), and enzymes that break down tissues to spread infection. Some bacteria, like *E. coli* or *Streptococcus*, use these factors to survive, grow, and cause diseases like food poisoning or strep throat.

2. Antifungal Activity

Molecular or Cellular Target/Pathway: Fungal ergosterol synthesis, cell wall integrity, mitotic spindle function, and mycelial growth pathway.

Mechanism: Cinnamon essential oil targets the ergosterol biosynthesis pathway in fungi, which is crucial for fungal cell membrane integrity. Ergosterol is an important molecule that helps make up the cell membranes of fungi, like how cholesterol works in animal cells. The process of making ergosterol involves several steps, starting with simpler building blocks like acetyl-CoA, which are turned into more complex molecules through different reactions. Eventually, these molecules combine to form ergosterol, which helps keep the cell membrane flexible and functioning properly. Inhibits cell wall integrity and mitotic spindle function by affecting beta-tubulin distribution, thereby causing cytotoxic effects. Certain substances can disrupt how cells maintain their structure and how they divide. First, it affects the cell wall, which is like a protective layer around the cell, making it weaker. Second, it interferes with the mitotic spindle, which is like a set of ropes that pull apart chromosomes during cell division. By changing where a protein called beta-tubulin is located, the substance makes it hard for the cells to divide correctly and stay strong. Cinnamon alters growth and reproduction in pathogenic fungi, such as *C. albicans*. Mycelial growth is how certain types of fungi, like mushrooms, spread and grow. It starts when tiny structures called spores land on a suitable surface, like soil or wood. The spores then begin to grow into thin, thread-like structures called hyphae, which spread out much like roots. These hyphae work together to form a larger network called mycelium. This mycelium helps the fungus absorb nutrients from its surroundings and can also help break down dead material, which is important for the ecosystem. When conditions are right, the mycelium can produce fruiting bodies, like mushrooms, that release more spores to continue the cycle.

3. Anti-inflammatory Properties

Molecular or Cellular Target/Pathway: Proinflammatory cytokines (IL-6, IL-1β, and TNF-α), NF-κB pathway, and NO production.

Mechanism: Cinnamon essential oil modulates the production of proinflammatory cytokines such as IL-6, IL-1β, and TNF-α. It also inhibits the NF-κB pathway, leading to decreased transcription of various inflammatory mediators. Cinnamon oil reduces induced nitric oxide production and other inflammatory markers while enhancing antioxidant enzyme activity to counteract oxidative stress.

4. Antioxidant Activity

Molecular or Cellular Target/Pathway: ROS scavenging and lipid peroxidation.

Mechanism: Cinnamon essential oil displays strong antioxidant properties that help in reducing oxidative stress and preventing damage caused by free radicals. Cinnamon scavenges free radicals and inhibits lipid peroxidation. Lipid peroxidation is a process that happens when fats in our body get damaged by things like stress, pollution, or unhealthy foods. When fats are damaged, they can turn into harmful molecules called free radicals, which can damage our cells and lead to inflammation and certain diseases—heart disease, cancer, and other health issues. It also enhances the activity of endogenous antioxidant enzymes and protects against oxidative damage in various tissues.

5. Antidiabetic Effects

Molecular or Cellular Target/Pathway: Alpha-glucosidase, alpha-amylase, PTB1B, and glucose metabolism pathways.

Mechanism: Cinnamon oil inhibits these enzymes to slow down carbohydrate digestion and absorption. Suppression of these enzymes slows the digestion of carbohydrates, reducing spikes in glucose and insulin after eating. It also inhibits this enzyme, leading to enhanced insulin sensitivity. PTP1B is an important enzyme in our bodies that helps regulate various cellular processes, especially those related to insulin and glucose metabolism. It works by removing phosphate groups from specific proteins, which can change the activity of those proteins. This process is crucial for controlling blood sugar levels and how our cells respond to insulin. Because PTP1B is involved in these processes, it has been studied for its role in obesity, diabetes, and other metabolic disorders. Lastly, cinnamon oil enhances glucose uptake in cells through modulation of insulin signaling pathways. By affecting these enzymes and pathways, cinnamon oil improves glucose tolerance, reduces blood glucose levels, and positively influences lipid profiles in diabetic models.

6. Cognitive and Neuroprotective Effects

Molecular or Cellular Target/Pathway: AChE cholinergic pathway, neuronal cell protection.

Mechanism: Cinnamon potently inhibits AChE, thereby increasing acetylcholine levels, which is crucial for learning and memory. Cinnamon essential oil enhances cognitive function by improving attentional processes and reducing neuron cell death.

7. Anticancer Activity

Molecular or Cellular Target/Pathway: Proapoptotic pathways and gene expression.

Mechanism: Cinnamon induces apoptosis in cancer cells through activation of proapoptotic pathways. Cinnamon impacts cell cycle transitions enhancing apoptosis in cancer cells of the prostate, breast, and other tissues. The main constituent in cinnamon bark oil, trans-cinnamaldehyde modulates gene expression related to inflammation, tissue remodeling, and directly inhibits tumor growth, triggering immune responses.

8. Respiratory Protection

Molecular or Cellular Target/Pathway: Viral virulence factors, disrupting virus release and infection spread.

Mechanism: Cinnamon inhibits neuraminidase and other viral proteins. Neuraminidase is an enzyme found on the surface of certain viruses, like the flu virus.

It helps the virus spread by cutting away sugars on the surface of infected cells, allowing new virus particles to escape and infect other cells. This makes neuraminidase a key target for antiviral medications because blocking it can help stop the virus from spreading in the body. Inhalation of cinnamon essential oil has been shown to increase survivability in influenza-infected models by inhibiting infection at multiple stages.

9. Moisture Retention and Wound Healing

Molecular or Cellular Target/Pathway: Fibroblast activity and collagen deposition.

Mechanism: Topical application accelerates healing processes through reduced inflammation, enhanced fibroblast activity, and increased angiogenesis. This involves several growth factors including VEGF and FGF-2 for angiogenesis and tissue repair. The FGF-2 pathway plays a crucial role in wound healing by promoting the growth and repair of tissues. It helps stimulate the formation of new blood vessels, which is important for delivering nutrients and oxygen to the healing area. FGF-2 also works alongside VEGF, another important protein that encourages blood vessel formation, making sure that the healing process goes smoothly.

10. Antimicrobial Resistance and Synergistic Effects

Molecular or Cellular Target/Pathway: Pathogen sensitivity to antibiotics, disruption of cell membranes and permeability.

Mechanism: Cinnamon enhances the effectiveness of antibiotics (e.g., gentamicin, clindamycin). It can restore sensitivity in resistant bacterial strains by damaging bacterial membranes or affecting permeability. The ability of cinnamon to enhance the effectiveness of established antibiotics makes it a promising candidate for controlling multiresistant infections.

CLARY SAGE

Clary sage essential oil, derived from the fragrant flowers and leaves of *Salvia sclarea*, boasts a diverse array of therapeutic properties that make it a valuable addition to natural health and wellness practices. Known for its potent anti-inflammatory effects, clary sage can inhibit prostaglandin production and modulate signaling through acetylcholine and oxytocin receptors, effectively addressing conditions such as dysmenorrhea and other inflammatory disorders. Its broad-spectrum antimicrobial activity disrupts bacterial cell membranes, enhancing the effectiveness of antibiotics against resistant strains, thus supporting wound healing and skin health. Moreover, clary sage has demonstrated endocrine and neurological benefits, influencing neurotransmitter levels to reduce stress and anxiety, and even exhibiting potential antidiabetic effects by regulating blood glucose metabolism. With its ability to promote wound healing and skin repair, as well as its cosmetic applications in addressing hyperpigmentation, clary sage essential oil is useful as a comprehensive solution for both physical and emotional well-being. Additionally, its insecticidal properties pave the way for natural pest control solutions, reinforcing its role in holistic health and sustainable living.

SUGGESTED THERAPEUTIC USE:

Inflammatory Conditions: Menstrual pain (dysmenorrhea), general inflammatory conditions

Infections: Wound infections, skin infections, antibiotic-resistant bacterial infections (in conjunction with antibiotics)

Neuroendocrine Dysregulation: Stress, anxiety, depression, mood regulation during menopausal transitions, social bonding issues

Metabolic Disorders: High blood glucose levels, insulin sensitivity issues

Wound and Skin Conditions: Wounds, skin repair and recovery

Mood and Stress: Anxiety, stress (especially in high-stress situations like labor or demanding work environments)

Insect Control and Repellence: Insect pest killing, repel pests

Skin Appearance: Hyperpigmentation (e.g., melasma, age spots)

MOLECULAR AND CELLULAR TARGETS, BIOLOGICAL PATHWAYS, AND MECHANISMS:

1. Anti-inflammatory Effects

Molecular or Cellular Target/Pathway: Prostaglandin pathway and release, AChE and oxytocin receptor activity, and NF-κB pathway.

Mechanism: Clary sage oil can inhibit prostaglandin production, which is involved in the inflammatory response. Compounds from clary sage modulate the signaling through these receptors, reducing uterine contractions and inflammation associated with dysmenorrhea. It also influences the NF-κB pathway, which regulates the expression of genes involved in inflammation. The essential oil has been shown to reduce inflammation through various mediators, suggesting its use in menstrual pain and other inflammatory conditions.

2. Antimicrobial Activity

Molecular or Cellular Target/Pathway: Pathogen cell wall/membrane integrity, growth inhibition, and pathogen antibiotic resistance.

Mechanism: Components of clary sage essential oil disrupt bacterial cell membranes, impacting the integrity and function of gram-positive and gram-negative bacteria. The oil possesses broad-spectrum antimicrobial properties, effectively inhibiting the growth of various pathogens, including *S. aureus*, *E. faecalis*, *C. albicans*, and others, which may be beneficial in wound healing and skin infections.

3. Endocrine and Neurological Effects

Molecular or Cellular Target/Pathway: Dopamine and serotonin receptors and activity, and oxytocin pathway.

Mechanism: Inhalation of clary sage may influence neurotransmitter levels associated with mood regulation, particularly in reducing stress and anxiety, and enhancing depressive symptom management. When inhaled, it increases oxytocin levels, impacting emotion, social behavior, and possibly mitigating some effects of aging. Oxytocin is a hormone that promotes social bonding, trust, and emotional connections between people, playing a significant role in human relationships and social interactions. It also has potential benefits for reducing stress and anxiety, contributing to overall emotional well-being and health. The essential oil exerts antidepressant-like effects, potentially improving mental well-being during menopausal transitions by influencing serotonergic activity and decreasing cortisol levels.

4. Antidiabetic Activity

Molecular or Cellular Target/Pathway: Alpha-glucosidase, insulin signal transduction pathway, and glucose-induced oxidative stress.

Mechanism: Clary sage influences insulin sensitivity and glucose metabolism in diabetic mice. Isolated compounds such as linalool are effective at reducing blood glucose levels by acting on metabolic pathways involved in glucose homeostasis. The whole oil is effective when sufficient linalool is present, but not as effective when linalyl acetate is the primary constituent. The oil potentially inhibits alpha-glucosidase, thus suppressing the release of glucose from carbohydrates in the small intestine. Research also suggested the oil possesses antioxidant activity, which may reduce glucose-induced oxidative stress. Lastly, the oil influences the insulin signal transduction pathway. The insulin signal transduction pathway is crucial for regulating blood glucose levels, and disruptions in this pathway are central to the development of diabetes. During normal insulin signaling, the pancreas releases insulin as glucose levels rise. Insulin binds to its receptors on the surface of cells, particularly in muscle, fat, and liver tissues, triggering a complex series of intracellular events, known as the insulin signaling pathway. Key proteins like insulin receptor substrates (IRS) and phosphoinositide 3-kinase (PI3K) are activated. This cascade ultimately leads to the translocation of glucose transporter 4 (GLUT4) to the cell surface. GLUT4 allows glucose to enter the cells from the bloodstream, lowering blood glucose levels. During diabetes, or disrupted insulin signaling, cells are resistant to insulin, meaning even when insulin is present, the signaling pathway is not properly activated. As a result, GLUT4 translocation is reduced, and glucose uptake is diminished. This leads to chronically elevated blood glucose levels (hyperglycemia).

5. Wound Healing and Skin Health

Molecular or Cellular Target/Pathway: Bacteria responsible for wound infections, fibroblast activity and collagen synthesis.

Mechanism: Clary sage's compounds may promote fibroblast activity and collagen synthesis, vital for skin repair and recovery. The oil targets bacteria responsible for wound infections, promoting healing and preventing infection. By inhibiting pathogenic bacteria and promoting cell proliferation, clary sage essential oil may enhance wound healing.

6. Anxiolytic and Sedative Properties

Molecular or Cellular Target/Pathway: HPA axis and dopaminergic pathway.

Mechanism: Inhalation or topical use of clary sage has been shown to reduce physiological signs of stress and anxiety, especially in situations like labor or in high-stress work environments. The oil's impact on cortisol suggests modulation of the hypothalamic-pituitary-adrenal (HPA) axis, a key stress response pathway. Clary sage influences brain chemicals, like dopamine, that regulate anxiety and mood. Dopamine is a neurotransmitter that plays a crucial role in regulating mood, motivation, and pleasure. Dysregulation of dopamine pathways has been implicated in the pathophysiology of depression.

7. Insecticidal and Repellent Effects

Molecular or Cellular Target/Pathway: Insect nervous system.

Mechanism: Clary sage oil components exhibit toxicity and repellent properties toward certain insects, affecting their nervous systems or behavior. The oil disrupts insect feeding and mating behaviors, providing potential applications in pest control.

8. Cosmetic and Dermatological Applications

Molecular or Cellular Target/Pathway: Tyrosinase activity.

Mechanism: Clary sage essential oil moderately inhibits the enzyme responsible for melanin production, which could be beneficial in cosmetic applications for skin hyperpigmentation. The potential to influence skin pigment production aids in developing therapeutic agents aimed at conditions like melasma or age spots.

CLOVE

Clove essential oil, extracted from the aromatic buds of *Syzygium aromaticum* (syn. *Eugenia caryophyllata*), is a powerful natural remedy renowned for its diverse therapeutic properties. It can also be extracted from the leaves, stems, and flowers. One of its most compelling attributes is its anticancer activity, with studies showing that clove oil induces apoptosis in several cancer cell lines, including breast, colorectal, and lung cancers. This is achieved through the activation of proapoptotic pathways and the selective reduction of ATP production in cancer cells, ultimately promoting programmed cell death. In addition to its anticancer effects, clove oil is celebrated for its broad-spectrum antimicrobial properties, disrupting bacterial membranes and inhibiting the growth of pathogens like *E. coli* and *S. aureus*, while also providing anti-inflammatory benefits by modulating proinflammatory cytokines and pathways. Furthermore, its neuroprotective effects, characterized by the inhibition of acetylcholinesterase and enhanced dopamine receptor activity, contribute to cognitive health. Alongside its antioxidant and gastroprotective properties, clove essential oil also supports immune function and displays antiparasitic and insecticidal capabilities, making it a versatile addition to health, wellness, and natural pest control solutions.

SUGGESTED THERAPEUTIC USE:

Cancer: Breast cancer, colorectal cancer, lung cancer

Infections: Bacterial infections, fungal infections, biofilm-related infections (chronic wound infections, catheter-associated UTIs, heart valve infections, chronic sinusitis, cystic fibrosis, dental plaque, periodontal disease, osteomyelitis, implant-associated infections, ventilator-associated pneumonia, conjunctivitis, dialysis-associated infections)

Neurological Conditions: Alzheimer's disease, mood disorders

Inflammatory Conditions: Arthritis, chronic inflammatory conditions (rheumatoid arthritis, IBD, psoriasis, COPD, asthma, lupus, eczema, gout, chronic sinusitis, fibromyalgia, chronic fatigue syndrome, periodontitis), conditions with excessive leukotriene production (asthma, seasonal allergies,

COPD, eosinophilic esophagitis, psoriasis, allergic conjunctivitis, lupus, cystic fibrosis, gout)

Organ and Tissue Protection: Oxidative stress-related conditions (cancer, IBD, diabetes, neurodegenerative conditions, cardiovascular disease, liver or kidney disease, metabolic syndrome, fibromyalgia, chronic fatigue syndrome, skin aging), cellular damage

Gastroprotection: Gastric ulcers, gut health issues

Immune Regulation: Infections, autoimmune conditions, weakened immune system

Parasites: Leishmaniasis, Chagas disease

Insect Control and Repellence: Mosquito infestations, lice infestations, insect pests

MOLECULAR AND CELLULAR TARGETS, BIOLOGICAL PATHWAYS, AND MECHANISMS:

1. Anticancer Activity

Molecular or Cellular Target/Pathway: Proapoptotic pathway, cellular energy metabolism, caspase activation, and gene expression.

Mechanism: Clove oil induces apoptosis in several cancer cell lines, including breast, colorectal, and lung cancer cells. Eugenol selectively reduces ATP production in cancer cells, impacting their energy availability. It induces caspase-3 activity, crucial for the apoptosis pathway. Caspase-3 is an essential executioner protease in the apoptotic (programmed cell death) pathway, and its activity is crucial for eliminating damaged or abnormal cells, including cancer cells. In cancer cells, decreased caspase-3 activity can lead to resistance to apoptosis, allowing these cells to survive and proliferate uncontrollably, while increased caspase-3 activity can induce apoptosis, helping to suppress tumor growth and progression. It also alters signaling pathways involved in cellular growth and survival, promoting programmed cell death in cancerous cells, likely by altering gene expression. Clove oil's anticancer effects are attributed to its ability to alter cellular metabolism and induce apoptotic processes in cancer cells, offering potential for cancer prevention and therapy.

2. Antimicrobial Activity

Molecular or Cellular Target/Pathway: Bacterial cell membrane integrity, pathogen enzyme activity, and quorum sensing.

Mechanism: Clove oil disrupts bacterial membranes, leading to cell lysis and death. Clove inhibits enzymes involved in bacterial growth and virulence, such as glucosyltransferases (GTFs) from *S. mutans*. GTFs are enzymes that play a key role in bacterial growth, especially in biofilm formation and pathogenesis. They help bacteria grow, stick to surfaces, resist attacks, and spread infections. Clove oil also inhibits biofilm formation in various bacteria and fungi, enhancing susceptibility to treatment. Clove essential oil has broad-spectrum antimicrobial activity against pathogens like *E. coli*, *S. aureus*, and *C. albicans*, effectively preventing infections and reducing biofilm production.

3. Neuroprotective Effects

Molecular or Cellular Target/Pathway: AChE and dopaminergic pathways.

Mechanism: Eugenol inhibits AChE, thereby increasing acetylcholine levels, which is beneficial in neurodegenerative diseases like Alzheimer's. Clove oil improves dopamine receptor sensitivity and activity, potentially ameliorating mood disorders. Dopamine receptor sensitivity and activity play a significant role in mood because dopamine is a "feel-good" neurotransmitter that helps control pleasure, motivation, and emotions. If dopamine receptors are highly sensitive and active, a person may feel more happy, motivated, and focused. But if the receptors become less sensitive (called downregulation), dopamine signals don't work as well, which can lead to low energy, lack of motivation, and even depression. On the other hand, too much dopamine activity in certain brain areas can cause anxiety, impulsive behavior, or even symptoms of schizophrenia. Keeping dopamine levels balanced helps maintain a stable and positive mood. By modulating neurotransmitter levels and providing neuroprotection, clove oil can help improve cognitive function and manage symptoms in various neurological conditions.

4. Anti-inflammatory Activity

Molecular or Cellular Target/Pathway: Proinflammatory cytokines (e.g., IL-6, TNF-α) and 5-LOX pathway.

Mechanism: Clove oil and eugenol decrease the levels of cytokines involved in inflammatory pathways (e.g., IL-6, TNF-α). It moderately inhibits the 5-LOX enzyme, which plays a crucial role in leukotriene production and inflammation. The anti-inflammatory effects observed when using clove oil can potentially reduce chronic inflammation, making it relevant for conditions such as arthritis.

5. Antioxidant Properties

Molecular or Cellular Target/Pathway: Free radical scavenging, endogenous antioxidant activity (GSTs, SOD).

Mechanism: Clove oil scavenges free radicals and inhibits oxidative stress markers. It enhances the activity of antioxidant enzymes (e.g., glutathione S-transferases and SOD). Clove essential oil effectively protects against oxidative stress, reducing lipid peroxidation and preventing cell damage.

6. Gastroprotective Benefits

Molecular or Cellular Target/Pathway: Gastric mucus production.

Mechanism: Clove oil stimulates mucus production, enhancing gastric mucosal defense. By promoting the production of the protective mucosal layer, clove oil helps prevent gastric ulcers and promote gut health.

7. Immunomodulatory Effects

Molecular or Cellular Target/Pathway: White blood cell proliferation and activity.

Mechanism: Clove enhances immune responses, providing immune support and reducing inflammation, making it potentially beneficial in infections and autoimmune conditions. It does so by increasing the proliferation and activity of immune cells and regulating the expression and activity of both proinflammatory (IL-1β, IL-6) and anti-inflammatory (IL-10) cytokines, restoring immune function in immunosuppressed models. This indicates that eugenol, the major component of clove, may exert its immunomodulatory/anti-inflammatory effects by suppressing NF-κB pathway. Clove oil significantly enhances both primary and secondary humoral immune responses, enhancing antibody production. This points to clove's ability to boost B-cell activity and antibody production. Restoration of cellular immune responses after immunosuppression suggests that clove affects T cell function.

8. Antiparasitic Activity

Molecular or Cellular Target/Pathway: Parasitic cellular structures.

Mechanism: Clove oil shows efficacy against the protozoans *Leishmania donovani* and *Trypanosoma cruzi*. *Leishmania donovani* is a protozoan parasite responsible for causing visceral leishmaniasis, which primarily affects internal organs such as the spleen, liver, and bone marrow. Symptoms include fever, weight loss, enlarged spleen or liver, and anemia. *Trypanosoma cruzi* is a protozoan parasite that causes Chagas disease, also known as American trypanosomiasis. This infection is primarily endemic to parts of Latin America but has also been reported in other regions due to human migration. The disease is mainly transmitted to humans through the feces of triatomine bugs (often referred to as kissing bugs). Chagas disease can present to two phases: 1) the acute phase, lasting from a few weeks to a couple of months and characterized by mild (fever, fatigue, body aches, rash, swelling at bite site) or no symptoms; and 2) the chronic phase, which is developed by about 30% of infected individuals and includes cardiac (arrhythmias, cardiomyopathy, and heart failure), gastrointestinal (megaesophagus, megacolon), and central nervous system problems.

9. Insecticidal and Repellent Properties

Molecular or Cellular Target/Pathway: Insect nervous systems.

Mechanism: The active compounds in clove disrupt the central nervous system of insects, leading to paralysis or death. Clove's insecticidal properties extend to various pests, including mosquitoes and lice, making it useful in pest control applications.

COPAIBA

Copaiba essential oil, derived from the resin of the *Copaifera* tree, is a powerhouse of therapeutic properties that has garnered attention for its myriad health benefits. Although obtained from many different *Copaifera* trees, the essential oil is most often rich in beta-caryophyllene, alpha-humulene, trans-Alpha-bergamotene, and other active sesquiterpenes. Indeed, it contains the highest levels of beta-caryophyllene of any known essential oil, with one published study showing a whopping 85.4%. Renowned for its potent anti-inflammatory action, copaiba oil effectively inhibits key enzymes like 5-lipoxygenase (5-LOX) and nitric oxide synthase (NOS), resulting in reduced production of proinflammatory mediators and cytokines. This modulation not only alleviates inflammation in tissues such as the skin and lungs but also contributes to its analgesic properties by interacting with pain-signaling pathways and opioid receptors. Beyond its anti-inflammatory and pain-relieving effects, copaiba is celebrated for its wound-healing capabilities, promoting angiogenesis and tissue regeneration through the activation of growth factors like VEGF. With broad-spectrum antimicrobial activity against various pathogens, antioxidant properties that combat oxidative stress, and neuroprotective effects beneficial for cognitive health, copaiba essential oil serves as a versatile natural remedy. It even shows promise in the management of endometriosis and anxiety reduction, while enhancing transdermal drug delivery and regulating lipid metabolism, making it an essential addition to holistic health practices.

SUGGESTED THERAPEUTIC USE:

Inflammatory Conditions: Skin inflammation, lung inflammation, intestinal inflammation

Pain: General pain, chronic pain

Wounds: Wounds, tissue damage

Infections: Oral infections, skin infections

Organ and Tissue Protection: Oxidative stress-related conditions (cancer, IBD, diabetes, neurodegenerative conditions, cardiovascular disease, liver or kidney disease, metabolic syndrome, fibromyalgia, chronic fatigue syndrome, skin aging), cellular damage, aging

Neurological Issues: Neuroinflammation, cognitive decline

Endometriosis: Endometriosis symptoms

Anxiety and Stress: Anxiety, stress

Transdermal Therapeutic Driver: Topical therapeutic agent absorption

Lipid Metabolism: Metabolic disorders, body composition issues

MOLECULAR AND CELLULAR TARGETS, BIOLOGICAL PATHWAYS, AND MECHANISMS:

1. Anti-inflammatory Action

Molecular or Cellular Target/Pathway: 5-LOX pathway, NOS signaling, JAK/STAT and MAPK.

Mechanism: Copaiba strongly inhibits the 5-LOX enzyme, which is involved in the conversion of arachidonic acid to leukotrienes, potent proinflammatory mediators. Inhibition leads to a decrease in leukotriene production, thereby reducing inflammation and associated symptoms. It also impacts the synthesis of nitric oxide (NOS signaling), a signaling molecule involved in inflammation. Copaiba upregulates the JAK/STAT and

MAPK signaling pathways, which are involved in immune response and cell signaling during inflammation. By modulating these targets and pathways, copaiba reduces the production of proinflammatory cytokines and mediators, leading to decreased inflammation in tissues (e.g., skin, lungs, and intestinal tissue).

2. Pain Relief (Analgesic Activity)

Molecular or Cellular Target/Pathway: Opioid receptors and pain-signaling pathways.

Mechanism: Since copaiba's antinociceptive effects are reversed by naloxone, this suggests copaiba interacts with opioid receptors. Copaiba modulates peripheral and central pain pathways, possibly through inhibition of pain neurotransmitters and inflammatory signals. The combined effects of reducing inflammation and directly acting on pain-signaling pathways contribute to analgesic properties.

3. Wound Healing and Angiogenesis

Molecular or Cellular Target/Pathway: Growth factors (e.g., VEGF), cytokines, and angiogenesis.

Mechanism: Copaiba essential oil promotes blood vessel formation and tissue regeneration likely by affecting VEGF. VEGF is a growth factor that plays a crucial role in angiogenesis, the process of forming new blood vessels from existing ones, which is essential for wound healing. VEGF also increases the permeability of blood vessels, allowing for easier transport of nutrients, fluids, and immune cells to the site of injury, which supports the healing process. Besides endothelial cells, VEGF can also influence the migration and proliferation of other cell types involved in wound healing, including fibroblasts and keratinocytes, further facilitating tissue repair. VEGF plays a role in modulating the inflammatory response, which is vital for the initial stages of wound healing. It helps recruit inflammatory cells to the wound site, which are necessary for clearing debris and pathogens. By promoting angiogenesis and cellular proliferation, VEGF indirectly supports the formation of the extracellular matrix, which is important for structural integrity and repair of the tissue. Copaiba oil accelerates wound healing through enhanced angiogenesis and reduced inflammation, facilitating tissue repair and regeneration.

4. Antimicrobial Activity

Molecular or Cellular Target/Pathway: Pathogen cell wall/membrane integrity.

Mechanism: Copaiba disrupts the structure and function of bacterial membranes. The oil's constituents cause damage to the membranes of bacteria and fungi, leading to their death or inhibition of growth. Copaiba displays activity against several oral pathogens and skin pathogens, contributing to its antiseptic properties.

5. Antioxidant Properties

Molecular or Cellular Target/Pathway: ROS and antioxidant defense pathways.

Mechanism: Copaiba protects cells from oxidative damage due to ROS. It activates various antioxidant pathways that reduce oxidative stress in tissues. By scavenging free radicals and inhibiting lipid peroxidation, copaiba oil mitigates oxidative damage in tissues, aiding in the protection against chronic diseases.

6. Neuroprotective Effects

Molecular or Cellular Target/Pathway: Neuroinflammatory mediators, promotion of neuritogenesis, and the PI3K/Akt/mTOR pathway.

Mechanism: Copaiba modulates the production of proinflammatory molecules in the central nervous system, reducing neuroinflammation. Through its high beta-caryophyllene content, copaiba also may promote neuritogenesis. Neuritogenesis is the process by which neurons (the cells in our brain and nervous system that transmit information) grow and develop their long, branch-like projections called neurites. These neurites can eventually become axons or dendrites, which are crucial for how neurons communicate with each other. Think of neurons as the messengers in your brain. They send signals to tell other parts of your body what to do—like move your hand or feel happy. When neurons are developing, they grow these long, thin extensions called neurites. It's similar to how a plant grows branches. As the neurites grow, some will turn into axons and others into dendrites. Axons are like the long wires that send signals away from the neuron to other neurons or muscles. Dendrites are like the receiving ends that take in signals from other neurons. As healthy branches (neurites) grow out, they create pathways for communication in the brain, helping us think, learn, and react to the world around us. It also positively regulates the PI3K/Akt/mTOR pathway in neuronal cells, which is crucial for cell metabolism and survival. Copaiba's neuroprotective effects may help mitigate neuroinflammation and promote healthy cognitive functions.

7. Management of Endometriosis

Molecular or Cellular Target/Pathway: Reproductive hormone pathways, proapoptotic pathway (endometrial cysts), anti-angiogenesis, CB2 activation, and inflammatory pathways.

Mechanism: Beta-caryophyllene triggers apoptosis in the luminal epithelium of endometrial cysts and endothelial cells of blood vessels. This proapoptotic activity can help reduce endometriosis. It has been shown to inhibit angiogenesis, the formation of new blood vessels. By limiting blood supply to endometrial cysts, beta-caryophyllene may impede their growth and survival. Activation of CB2 receptors by beta-caryophyllene can modulate inflammation and immune responses, which may contribute to the reduction of endometriosis-related symptoms. Increased plasma levels of testosterone and estradiol suggests that beta-caryophyllene balances endocrine system function and therefore, hormones. Beta-caryophyllene is known to promote estrogen and testosterone balance in women. By modulating inflammatory processes, regulating endometrial tissue growth, and potentially affecting hormonal signaling and balance, copaiba reduces the severity of endometriosis in animal models.

8. Anxiety and Stress Reduction

Molecular or Cellular Target/Pathway: Serotonin and dopamine levels.

Mechanism: Inhalation of copaiba essential oil may affect neurotransmitter levels like dopamine and serotonin, facilitating anxiety management and improved mood. Some studies indicate that copaiba may have calming effects, reducing physiological stress responses and improving cognitive function.

9. Enhancement of Transdermal Drug Delivery

Molecular or Cellular Target/Pathway: Temporarily and reversibly increases skin permeability.

Mechanism: Copaiba can alter skin permeability, enhancing the delivery of lipophilic substances. This activity improves penetration of topical medications (e.g., celecoxib) and other essential oils through the skin barrier. As a natural penetration enhancer, copaiba essential oil facilitates the absorption of therapeutics, improving healing effects.

10. Regulation of Lipid Metabolism

Molecular or Cellular Target/Pathway: Fatty acid synthesis, oxidation enzymes, and lipid metabolism pathways.

Mechanism: Copaiba essential oil influences lipid processing through metabolic pathways. Fatty acid synthesis and oxidation enzymes help regulate lipid metabolism by controlling how the body stores and uses fat for energy. Fatty acid synthesis enzymes, like acetyl-CoA carboxylase (ACC) and fatty acid synthase (FAS), help create new fat molecules when the body has extra energy, storing them for later use. On the other hand, fatty acid oxidation enzymes, like carnitine palmitoyltransferase (CPT) and acyl-CoA dehydrogenase, break down stored fat into energy when the body needs fuel, like during exercise or fasting. These two processes must stay balanced—too much fat storage can lead to obesity, while too much fat breakdown can cause energy shortages. The oil may help regulate fat storage and mobilization in tissues. By modulating lipid metabolism, copaiba may help in managing body composition and reducing inflammation linked to metabolic disorders.

EUCALYPTUS

Eucalyptus essential oil, most commonly derived from the leaves of *Eucalyptus globulus* and *Eucalyptus radiata*, is a remarkable natural remedy known for its extensive health benefits, particularly in promoting respiratory health. Key components like 1,8-cineole work synergistically to reduce inflammation and mucus production in airway epithelial cells while modulating Toll-like Receptor 4 (TLR4) activity, thus preventing excessive inflammatory responses often associated with chronic respiratory conditions such as asthma and COPD. Beyond its respiratory benefits, eucalyptus oil boasts potent antimicrobial properties, effectively disrupting the membranes of various bacteria and fungi, making it a formidable natural antibiotic against drug-resistant pathogens. Its analgesic effects are equally noteworthy; by influencing opioid receptors and reducing proinflammatory substances, eucalyptus provides relief from pain, including conditions like arthritis. Furthermore, its neuroprotective and antioxidant actions enhance cognitive function and combat oxidative stress, while also supporting skin health through improved wound healing and antibacterial activity. With its versatile capabilities that extend to insecticidal properties and gastrointestinal benefits, eucalyptus essential oil stands as a vital addition to holistic health practices, championing both wellness and preventive care.

SUGGESTED THERAPEUTIC USE:

Respiratory Conditions: Asthma, COPD, respiratory infections, airway inflammation, mucus congestion

Infections: Bacterial infections, fungal infections, drug-resistant infections, biofilm-related infections (chronic wound infections, catheter-associated UTIs, heart valve infections, chronic sinusitis, cystic fibrosis, dental plaque, periodontal disease, osteomyelitis, conjunctivitis, implant-associated infections, ventilator-associated pneumonia, dialysis-associated infections)

Pain: Arthritis pain, general pain

Neurological and Mood Conditions: Anxiety, stress, mood disorders, cognitive decline, stroke recovery, diabetes-related neurological complications

Inflammatory Conditions: Chronic inflammatory conditions (rheumatoid arthritis, IBD, psoriasis, COPD, asthma, lupus, eczema, gout, chronic sinusitis, fibromyalgia, chronic fatigue syndrome, periodontitis), conditions with excessive leukotriene production (asthma, seasonal allergies, COPD, eosinophilic esophagitis, psoriasis, allergic conjunctivitis, lupus, cystic fibrosis, gout)

Skin Infections and Wounds: Wounds, skin infections, slow-healing wounds

Organ and Tissue Protection: Oxidative stress-related conditions (cancer, IBD, diabetes, neurodegenerative conditions, cardiovascular disease, liver or kidney disease, metabolic syndrome, fibromyalgia, chronic fatigue syndrome, skin aging), cellular damage

Insect Control and Repellency: Mosquito infestations, cockroach infestations, insect pests

Gastroprotection: Gut dysbiosis, gut inflammation

Viral Infections: Respiratory viral infections (e.g., H1N1, SARS-CoV-2)

MOLECULAR AND CELLULAR TARGETS, BIOLOGICAL PATHWAYS, AND MECHANISMS:

1. Respiratory Health

Molecular or Cellular Target/Pathway: TLR4 activity, airway inflammation, and airway mucus production.

Mechanism: Eucalyptus oil acts as an anti-inflammatory agent in the respiratory tract, alleviating symptoms associated with respiratory infections and reducing airway hyperreactivity. One mechanism it uses to do so is modulating TLR4 activity to prevent excessive inflammatory responses. TLR4 is expressed on various immune and airway cells, including epithelial cells and lung macrophages. When TLR4 detects specific molecules, such as lipopolysaccharides (from Gram-negative bacteria) or certain allergens, it activates signaling pathways that lead to the production of proinflammatory cytokines and chemokines. These substances help recruit more immune cells to the site of inflammation and amplify the inflammatory response. In conditions like asthma, chronic obstructive pulmonary disease (COPD), and other respiratory infections, TLR4 activation can contribute to excessive airway inflammation, leading to symptoms such as wheezing, coughing, and difficulty breathing. Additionally, chronic activation of TLR4 can promote tissue remodeling and airway hyperresponsiveness, worsening these conditions over time. 1,8-cineole and other components reduce inflammation and mucus production in the airways. Reduces inflammation caused by lipopolysaccharides, which can result in chronic respiratory conditions like asthma and COPD.

2. Antimicrobial Activity

Molecular or Cellular Target/Pathway: Pathogen cell wall/membranes and biofilm formation.

Mechanism: Eucalyptus essential oil exhibits broad-spectrum antimicrobial activity, acting as a natural antibiotic that can combat drug-resistant bacteria and fungal infections. Eucalyptus disrupts pathogen cell wall/membrane integrity in various pathogens. It also reduces the adherence of bacteria to surfaces, preventing biofilm development that makes pathogens more resistant to antibiotics.

3. Pain Relief and Analgesic Properties

Molecular or Cellular Target/Pathway: Opioid receptors, proinflammatory cytokines, triggering receptor expressed on myeloid cells-1 (TREM-1), MKP-1 phosphatase, hTRPM8, P2X3, and prostaglandins.

Mechanism: Your body's pain system is a complex network involving multiple mechanisms and pathways. Eucalyptus essential oil provides relief from conditions such as arthritis by acting on multiple pain pathways and reducing sensitivity to pain. Opioid receptors are like special locks that, when picked by substances that interact with them, block pain signals from reaching your brain. Eucalyptus influences central and peripheral pain pathways through these receptors to provide analgesic effects. Proinflammatory cytokines are like tiny alarm bells that your body releases when it's injured or infected, causing inflammation and pain. Eucalyptus reduces the activity of proinflammatory compounds that contribute to pain. TREM-1 is like a master switch for those alarm bells, amplifying the pain response when triggered. 1,8-cineole diminishes TREM-1 activity. MKP-1 phosphatase acts like a dimmer switch, turning down the intensity of the inflammatory alarm bells. Both eucalyptus and 1,8-cineole affect MKP-1 phosphatase to reduce pain. hTRPM8 is a cold sensor that, when activated by substances like menthol or eucalyptus, can override some pain signals. P2X3 is a receptor that helps transmit pain signals, especially those related to nerve pain, so blocking it can reduce that type of pain. Eucalyptus essential oil and 1,8-cineole are the activators of hTRPM8, the antagonist of hTRPA1, and the inhibitor of the P2X3 receptor, which reduces pain and inflammation. Finally, prostaglandins are like chemical messengers that amplify pain signals and contribute to inflammation, making you more sensitive to pain. It is believed that eucalyptus interferes with the formation of prostaglandins.

4. Neuroprotective Effects

Molecular or Cellular Target/Pathway: Dopamine, GABA receptors, SIRT1, and Nrf2

Mechanism: Eucalyptus essential oil may provide neuroprotection during cerebral events, such as strokes, through its ability to reduce oxidative stress and inflammation. By inhaling eucalyptus oil, there is an observed increase in dopamine levels in the brain, which can positively affect mood and cognitive function. Moreover, eucalyptus oil promotes the release of GABA, contributing to a sense of calmness and reduced anxiety, further supporting neurological health. Importantly, eucalyptus oil activates the SIRT1 and Nrf2 pathways, enhancing cellular defense mechanisms against oxidative damage. This activation is

particularly beneficial in counteracting diabetes-related complications, as it may protect neurons from injury and promote overall brain health. Through these multifaceted actions, eucalyptus essential oil serves as a potential adjunct therapy for neuroprotection during and after ischemic events.

5. Anti-inflammatory Effects

Molecular or Cellular Target/Pathway: Proinflammatory cytokines (IL-4, TNF-α), NO, and NF-κB pathways.

Mechanism: Eucalyptus exhibits acute anti-inflammatory properties that can alleviate conditions influenced by chronic inflammation. It downregulates cytokines like IL-4 and TNF-α involved in inflammation responses. It also scavenges NO free radicals to prevent abnormal inflammation. Finally, eucalyptus may inhibit the NF-κB pathway, which triggers inflammatory responses at the transcriptional level.

6. Skin Health and Wound Healing

Molecular or Cellular Target/Pathway: Collagen synthesis and wound pathogens.

Mechanism: The application of eucalyptus essential oil boosts wound healing through antimicrobial properties and stimulation of tissue regeneration. Specifically, it enhances collagen synthesis at the wound site and targets common pathogens in wound infections, promoting effective healing.

7. Antioxidant Activity

Molecular or Cellular Target/Pathway: Oxidative stress.

Pathway: Eucalyptus prevents lipid peroxidation and scavenges free radicals, thus protecting cells from oxidative damage. It protects cells by activating antioxidant pathways that restore enzymatic activity and combat the harmful effects of oxidative stress. Eucalyptus may enhance the activities of endogenous antioxidant enzymes such as CAT and SOD.

8. Insecticidal and Repellent Properties

Mechanisms: Eucalyptus essential oil demonstrates significant efficacy as a natural insecticide and repellent, offering a potential alternative to conventional chemical insecticides. It does so by disrupting normal function leading to paralysis or death in pests like mosquitoes and cockroaches.

9. Gastroprotective Properties

Molecular or Cellular Target/Pathway: Gut microbiome.

Mechanism: Eucalyptus improves gut health by fostering beneficial bacteria and modulating inflammatory responses. It enhances the biodiversity of gut microflora which is essential for metabolic health and regulation of mood through the gut-brain axis.

10. Antiviral Activity

Molecular or Cellular Target/Pathway: Viral membrane disruption.

Mechanisms: By disrupting the lipid bilayer of viruses, eucalyptus oil hinders their ability to infect host cells. The oil exhibits potent activity against respiratory viruses like H1N1 and SARS-CoV-2, indicating potential applicability in respiratory infections.

FENNEL (SWEET/BITTER)

Fennel essential oil, extracted from the aromatic seeds of the fennel plant (*Foeniculum vulgare*), is a dynamic natural remedy revered for its diverse therapeutic properties. Known for its potent antifungal activity, fennel oil disrupts the integrity of fungal cell membranes, making it effective against troublesome species such as *C. albicans* and *Trichophyton*. Additionally, this essential oil showcases significant antimicrobial properties, demonstrating efficacy against various bacterial strains, including *K. pneumoniae* and *S. aureus*, by inhibiting biofilm formation and disrupting bacterial membranes. Beyond its antimicrobial capabilities, fennel oil possesses anxiolytic and sedative effects, acting on the central nervous system to alleviate anxiety while enhancing mood—largely attributed to its constituent trans-Anethole. Its anticancer potential is equally compelling, as fennel oil has been shown to induce apoptosis in breast and ovarian cancer cells, suggesting a promising avenue for therapeutic applications in cancer treatment. With anti-inflammatory, gastroprotective, and cardiovascular benefits, along with its neuroprotective and insecticidal properties, fennel essential oil is a multifaceted ally in promoting health and well-being across a range of conditions.

SUGGESTED THERAPEUTIC USE:

Fungal Infections: Fungal infections, dermatophyte infections, skin infections, nail infections, hair infections

Bacterial Infections: Bacterial infections, biofilm-related infections (chronic wound infections, catheter-associated UTIs, heart valve infections, chronic sinusitis, cystic fibrosis, dental plaque, periodontal disease, osteomyelitis, conjunctivitis, implant-associated infections, ventilator-associated pneumonia, dialysis-associated infections)

Mood and Sleep Problems: Anxiety, stress, insomnia (at higher doses)

Cancer: Ovarian cancer, breast cancer

Inflammatory Conditions: Inflammatory conditions, gastrointestinal inflammation, colitis

Gastroprotection: Gastric ulcers, gastrointestinal issues

Blood Pressure: High blood pressure, cardiovascular health issues

Organ and Tissue Protection: Oxidative stress-related conditions (cancer, IBD, diabetes, neurodegenerative conditions, cardiovascular disease, liver or kidney disease, metabolic syndrome, fibromyalgia, chronic fatigue syndrome, skin aging), cellular damage

Hormone Regulation and Female Reproductive Conditions: Menstrual pain, irregular periods, menstrual disorders

Neurological Issues: Neurodegenerative disorders, anxiety, depression, cognitive decline

Pest Control: Insect infestations (e.g., red flour beetle, head lice)

MOLECULAR AND CELLULAR TARGETS, BIOLOGICAL PATHWAYS, AND MECHANISMS:

1. Antifungal Activity

Molecular or Cellular Target/Pathway: Fungi cell membranes.

Mechanism: The essential oil demonstrates broad-spectrum antifungal activity, potentially serving as a therapeutic agent against dermatophyte infections affecting the skin, hair, and nails. Fennel oil disrupts the integrity of fungal cells, and inhibits mycelial growth and spore germination, preventing fungal infection and spread.

2. Antimicrobial Properties

Molecular or Cellular Target/Pathway: Bacterial cell membranes and biofilm formation.

Mechanism: Fennel essential oil demonstrates antibacterial activity via membrane disruption and potential inhibition of bacterial growth and metabolism. It also reduces biofilm formation, enhancing susceptibility to antimicrobial treatments.

3. Anxiolytic and Sedative Properties

Molecular or Cellular Target/Pathway: Serotonin, dopamine, and adrenaline, and sympathetic nervous system activity.

Mechanism: The presence of constituents like trans-Anethole may modulate anxiety levels by optimizing neurotransmitter systems in the brain. Fennel oil influences neurotransmitter systems, particularly regarding serotonin and dopamine. It increases adrenaline levels and sympathetic activity, promoting alertness at low doses, while higher doses exhibit sedative effects.

4. Anticancer Activity

Molecular or Cellular Target/Pathway: Cell signaling pathways and BCL-2 family proteins.

Mechanism: The oil promotes cell death in cancerous cells while being less damaging to normal cells, suggesting its therapeutic potential for cancer. Its major constituent, trans-Anethole, has been shown to induce apoptosis in ovarian and breast cancer cells, increasing proapoptotic proteins while decreasing antiapoptotic signals, particularly BCL-2 family proteins. BCL-2 family proteins are a group of proteins crucial for regulating cell death, particularly a process called apoptosis, which is the body's way of eliminating damaged or unnecessary cells. Think of these proteins as a balance system: some, like BCL-2 and BCL-XL, act as life preservers that help keep cells alive, while others, like Bax and Bak, are like signals that trigger cell death when needed. In cancer, this balance can be disrupted; for instance, cancer cells may overproduce the life-preserving proteins, allowing them to survive longer than they should. This helps tumors grow and resist treatments like chemotherapy, which aim to kill cancer cells.

5. Anti-inflammatory Properties

Molecular or Cellular Target/Pathway: NF-κB and oxidative stress.

Mechanism: Fennel oil impacts mediators such as NF-κB, which is crucial in inflammatory responses. It also enhances antioxidant pathways, reducing overall inflammation and protecting tissues. By inhibiting proinflammatory signals, fennel essential oil may be beneficial in treating inflammatory conditions, including gastrointestinal diseases and colitis.

6. Gastroprotective Effects

Molecular or Cellular Target/Pathway: COX, LOX, maintenance of blood supply to the stomach lining, oxidation, and inflammation.

Mechanisms: Fennel oil and trans-Anethole protect against ethanol-induced gastric lesions. Their antioxidant and anti-inflammatory properties help maintain mucosal integrity and promote healing in the gastrointestinal tract. Fennel helps protect the stomach by inhibiting enzymes like COX, which play a role in producing substances that can cause inflammation and discomfort in the stomach. It also blocks another group of enzymes called LOX, which normally break down fatty acids in a way that can lead to irritation. Furthermore, fennel helps prevent platelets—tiny blood cells that clump together to form clots—from over-responding to issues in the stomach lining. This is important because it helps maintain a healthy blood supply to the stomach's protective lining, ensuring it stays healthy and functions well.

7. Blood Pressure Regulation and Vasorelaxation

Molecular or Cellular Target/Pathway: Calcium channels, potassium channels, and vasodilation.

Mechanism: Fennel essential oil increases calcium ion entry and affects potassium channels, influencing vascular smooth muscle relaxation. The vasorelaxing effects improve blood vessel function and may contribute to cardiovascular health by preventing excessive clotting and promoting blood flow.

8. Antioxidant Activity

Molecular or Cellular Target/Pathway: ROS and endogenous antioxidants (SOD, CAT).

Pathway: Fennel oil may protect cells from oxidative stress, contributing to overall health and longevity. It scavenges free radicals and reduces oxidative damage, while simultaneously increasing the activity of protective enzymes such as SOD and CAT, thus promoting cellular health.

9. Hormonal Regulation and Female Reproductive Health

Molecular or Cellular Target/Pathway: Estrogen receptors and hormonal pathways.

Mechanism: Fennel oil may influence menstrual health by interacting with hormonal pathways. Fennel contains anethole, a compound with structural similarity to estrogen. This allows it to bind to estrogen receptors and exert estrogen-like effects in the body. It also reduces uterine contractions induced by oxytocin and PGE2 in laboratory settings. Fennel influences female reproductive health through multiple pathways, including estrogenic activity, uterine muscle relaxation, and prostaglandin inhibition. It has been shown to relieve menstrual pain, regulate periods, and potentially alleviate symptoms associated with menstrual disorders.

10. Neuroprotection

Molecular or Cellular Target/Pathway: Currently unknown targets and pathways.

Mechanism: The mechanisms of action of fennel oil are not fully understood, but its antioxidant activity may protect against neurodegeneration and influences neurotransmitter levels (dopamine, serotonin). Other potential mechanisms include anxiolytic and antidepressant effects, inhibition of AChE, and reduced brain inflammation.

11. Pest Control and Insecticidal Activity

Molecular or Cellular Target/Pathway: Insect nervous systems.

Mechanism: The volatile compounds in fennel disrupt insect physiology, including pests such as the red flour beetle and head lice. Exhibits notable insecticidal and acaricidal properties, making it useful as a natural pest control agent.

FIR (BALSAM/SIBERIAN)

The essential oils of balsam fir (*Abies balsamea*) and Siberian fir (*A. sibirica*) essential oils have long been revered for their therapeutic potential, spanning diverse applications from traditional remedies to modern scientific inquiry. Recent studies shed light on the multifaceted mechanisms underlying their efficacy. Balsam fir essential oil exhibits potent anticancer properties, selectively targeting malignant cells across various cancer types, including breast, prostate, lung, colon, and melanoma. This activity is primarily attributed to the oil's ability to disrupt cellular redox balance by depleting glutathione and inducing reactive oxygen species (ROS) production. Furthermore, components of balsam fir demonstrate antimicrobial activity against *S. aureus*, and the whole-needle oil, along with beta-pinene, shows insecticidal efficacy against blacklegged ticks, with enhanced activity at lower temperatures. Synergistic effects with clothianidin against mosquitoes have also been observed. Siberian fir essential oil, in contrast, appears to primarily interact with neurological and enzymatic pathways. Although these two oils are not as well researched as others yet, clinical experience shows they are highly valuable oils for a variety of conditions including pain and inflammation.

SUGGESTED THERAPEUTIC USE:

Cancer: Breast, prostate, lung, colon, and melanoma

Cognitive Decline: Memory, cognitive function, dementia, Alzheimer's disease

Mental Health: Anxiety, stress

Infections: Bacterial, fungal, oral candidiasis

Insect-borne Diseases: Tick- and mosquito-borne diseases

Organ and Tissue Protection: Oxidative stress-related conditions (cancer, IBD, diabetes, neurodegenerative conditions, cardiovascular disease, liver or kidney disease, metabolic syndrome, fibromyalgia, chronic fatigue syndrome, skin aging), aging, chronic diseases

Respiratory Issues: Cough, congestion

MOLECULAR AND CELLULAR TARGETS, BIOLOGICAL PATHWAYS, AND MECHANISMS

1. Anticancer Activity (Balsam Fir)

Molecular or Cellular Target/Pathway: Depletion of cellular glutathione, increased (ROS) production.

Mechanisms: Induction of oxidative stress in cancer cells, leading to cellular damage and apoptosis. Disruption of cellular redox homeostasis. Cellular redox homeostasis plays a critical role in cancer prevention and treatment due to its influence on oxidative stress, cell survival, and apoptosis. Normal cells maintain redox balance through antioxidants and redox enzymes. Cancer cells often exploit redox imbalance, increasing reactive oxygen species (ROS) for proliferation while upregulating antioxidant defenses to avoid excessive oxidative damage. Targeting this imbalance can selectively induce oxidative stress in cancer cells, leading to cell death. Certain compounds, like ascorbate (high-dose vitamin C), function as pro-oxidants in cancer cells, selectively increasing oxidative stress to kill them. Inhibiting cancer cells' antioxidant defenses (e.g., GDH, SOD) sensitizes them to oxidative damage and enhances treatment efficacy. Many natural anticancer agents, including polyphenols, flavonoids, and essential oils, modulate redox homeostasis by either promoting ROS accumulation or restoring normal redox signaling in cancer cells. By strategically modulating redox homeostasis—either by increasing oxidative stress in cancer cells or restoring normal redox function—anticancer strategies can selectively target tumor cells while preserving healthy ones. Alpha-humulene is considered the primary active compound.

2. Antimicrobial Activity (Balsam Fir)

Molecular or Cellular Target/Pathway: Bacterial cell walls/membranes.

Mechanisms: Balsam fir disrupts bacterial cell membrane integrity. It also inhibits bacterial growth.

3. Insecticidal Activity (Balsam Fir)

Molecular or Cellular Target/Pathway: Neurological targets in blacklegged ticks (*Ixodes scapularis*) and mosquitoes.

Mechanisms: The oil possesses neurotoxic effects, leading to paralysis and death. It shows synergistic effects with insecticides like clothianidin and has a temperature-dependent efficacy.

4. Relaxation and Cognitive Effects (Siberian Fir)

Molecular or Cellular Target/Pathway: Brainwaves, AChE.

Mechanisms: Siberian fir modulates neurotransmitter activity, promoting relaxation and reducing arousal. Clinical research shows that the oil slows heart rate and increases theta brainwaves, suggesting meditative thought and decreased arousal. It also preserves acetylcholine levels, potentially enhancing cognitive function and memory.

5. Antioxidant Activity (Siberian Fir)

Molecular or Cellular Target/Pathway: Endogenous antioxidant activity (CAT, glutathione reductase).

Mechanisms: The oil enhances cellular antioxidant defenses and the activity of endogenous antioxidants (CAT, GR), reducing oxidative stress.

6. Antifungal Activity (Siberian Fir)

Molecular or Cellular Target/Pathway: Fungal cell walls/membranes, growth inhibition.

Mechanisms: Siberian fir oil disrupts fungal cell membrane integrity and inhibits their growth.

7. Larvicidal Activity (Siberian Fir)

Molecular or Cellular Target/Pathway: Direct toxicity to *Ae. aegypti* larvae.

Mechanisms: Siberian fir is directly toxic to mosquito larvae, leading to larval death.

FRANKINCENSE

Frankincense essential oil, often referred to as "the king of essential oils," emanates from the resin of the *Boswellia* species (most commonly *B. carterii, B. sacra, B. frereana,* and *B. papyrifera*) and is celebrated for its profound therapeutic properties. Its anticancer activity is particularly noteworthy, as frankincense has demonstrated the ability to induce apoptosis in various cancer cell lines, including bladder, pancreatic, and breast cancers, through mechanisms such as oxidative stress and modulation of critical signaling pathways like AMPK/mTOR. This valuable oil also showcases robust anti-inflammatory effects, effectively inhibiting proinflammatory cytokines and protecting healthy cartilage, making it beneficial for managing arthritis and other inflammatory conditions. Beyond its anti-inflammatory and anticancer attributes, frankincense possesses impressive antimicrobial properties that disrupt the membranes of both bacterial and fungal cells, thereby aiding in the fight against infections. Additionally, its immunomodulatory effects enhance immune responses while promoting tolerance, offering potential benefits for autoimmune diseases and allergies. With contributions to skin healing, pain relief, cardiovascular protection, and neuroprotection, frankincense essential oil holds a revered place in holistic health, promising a multitude of benefits for those who seek its natural power.

SUGGESTED THERAPEUTIC USE:

Cancer: Bladder cancer, pancreatic cancer, breast cancer, colon cancer

Inflammatory Conditions and Damage: Arthritis, inflammation-related tissue damage

Infections: Bacterial infections, fungal infections

Immune Balance: Autoimmune diseases, allergies

Skin Aging and Damage: Wrinkles, UV damage, wounds, skin aging

Pain: Chronic pain, cancer-related pain, neurogenic pain, inflammatory pain

Cardiovascular Issues: Myocardial infarction, ischemic conditions, heart damage

Neurological Disorders: Neurodegenerative disorders, anxiety, stress responses, cognitive decline

Ulcerative Colitis and Gastrointestinal Conditions: Ulcerative colitis, gastrointestinal inflammation, digestive conditions

Leishmaniasis: Leishmaniasis

MOLECULAR AND CELLULAR TARGETS, BIOLOGICAL PATHWAYS, AND MECHANISMS:

1. **Anticancer Activity**

Molecular or Cellular Target/Pathway: Cancer cell proapoptotic pathway, delta-catenin signaling, and AMPK/mTOR signaling.

Mechanism: Frankincense oil induces cell death in various cancer cell lines (e.g., bladder, pancreatic, and breast cancer) by inducing oxidative stress and apoptosis. Frankincense (*B. sacra*) reduces the proliferation of colon cancer cells through this specific molecular pathway. Delta-catenin is a protein that plays a role in cellular communication. Specifically, it's involved in pathways that control how cells grow and stick together. In healthy cells, delta-catenin helps keep everything in balance. However, in cancer, especially colon cancer, delta-catenin's signals may be abnormal, leading to cells growing too much or not sticking together properly. This uncontrolled growth and spread is what makes cancer dangerous. So, when delta-catenin signaling is irregular, it can contribute to the development and progression of colon cancer by disrupting normal cell behavior. Frankincense modulates the AMPK/mTOR signaling pathway, influencing cell growth and metabolic processes critical for cancer progression. The essential oil activates apoptotic pathways while selectively targeting cancer cells, suggesting its potential as a complementary therapeutic agent in oncology.

2. Anti-inflammatory Effects

Molecular or Cellular Target/Pathway: Proinflammatory cytokines, NO, NF-κB, and chondrocyte production.

Mechanism: Frankincense inhibits the production of interleukins (e.g., IL-1), nitric oxide, and NF-κB, which are involved in inflammatory responses. It protects healthy cartilage cells (chondrocytes) from breakdown due to inflammatory mediators. The essential oil's anti-inflammatory properties can mitigate conditions such as arthritis, thus reducing inflammation-related tissue damage.

3. Antimicrobial Properties

Molecular or Cellular Target/Pathway: Microbial cell membrane disruption.

Mechanisms: Frankincense inhibits various Gram-positive and Gram-negative bacteria and fungi by disrupting cell membrane integrity and metabolic pathways in bacteria and fungi. The variation in antimicrobial efficacy is linked to the specific composition of different frankincense oils, influencing their effectiveness.

4. Immunomodulatory Effects

Molecular or Cellular Target/Pathway: Lymphocyte and dendritic cell function and immune signaling pathways.

Mechanisms: Frankincense stimulates lymphocyte activation and modulates the differentiation of monocytes into immature dendritic cells, promoting immune tolerance. Immune tolerance is the body's ability to recognize and accept its own cells and harmless substances while still defending against harmful invaders like viruses and bacteria. Signal transduction pathways for immune modulation enable frankincense to enhance immune responses while promoting tolerance to prevent autoimmune reactions. The immune system relies on signal transduction pathways—a series of molecular switches inside cells—to control how it responds to threats and maintains balance. When a harmful invader, like a virus, is detected, these pathways activate immune defenses, triggering cells to attack and clear the infection. At the same time, other pathways act as brakes, preventing the immune system from mistakenly attacking the body's own cells, which could lead to autoimmune diseases like lupus or multiple sclerosis. Specialized immune cells, such as regulatory T cells (Tregs), use these signals to suppress unnecessary immune responses and promote immune tolerance, ensuring that the immune system reacts only when truly needed and stays in balance to protect overall health. These properties make frankincense beneficial for conditions such as autoimmune and autoinflammatory diseases and allergies by balancing immune responses.

5. Skin Aging and Healing

Molecular or Cellular Target/Pathway: Skin cells, extracellular matrix components.

Mechanisms: Frankincense enhances skin's antioxidant capacity, preserves epidermal thickness, and stimulates procollagen I production. Procollagen I is essentially the precursor to collagen type I, which is the most abundant type of collagen in the body, and it plays a critical role in the health and structure of your skin. It is fundamental for maintaining healthy, youthful-looking skin by providing the raw material necessary for collagen production. Enhances wound-healing processes and mitigates wrinkles and signs of skin aging due to UV exposure.

6. Analgesic Effects and Pain Relief

Molecular or Cellular Target/Pathway: COX and LOX pathways, neurogenic pain pathways, and peripheral pain mediators in local tissues.

Mechanisms: Frankincense essential oil reduced pain in both the early (neurogenic) and late (inflammatory) phases, indicating that it affects both nerve signaling and immune-mediated inflammation. The oil likely inhibits prostaglandin synthesis, similar to aspirin, reducing inflammation and pain. The oil may suppress leukotriene production and prostaglandin synthesis, decreasing inflammatory pain. Frankincense essential oil reduces pain perception by interacting with local pain mediators (e.g., prostaglandins). Topical application has been shown to reduce stress-induced pain, possibly by influencing cortisol levels and modulating neurotransmitter levels. Frankincense reduces neuronal hyperactivity to modulate pain signals at the spinal level and multiple proinflammatory mediators, and modulates pain-signaling pathways to promote pain relief. The analgesic effect enhances the quality of life in patients with chronic pain conditions, including cancer-related pain.

7. Cardiovascular Protection

Molecular or Cellular Target/Pathway: Cardiac tissue protection against oxidative stress.

Mechanisms: Frankincense provides protection during ischemic conditions, potentially influencing factors that contribute to heart damage. Through its antioxidant properties, it helps in attenuating damage caused during myocardial infarctions (heart attacks).

8. Neuroprotective Properties

Molecular or Cellular Target/Pathway: Neurotransmitter regulation.

Mechanisms: Frankincense oil influences levels of serotonin and norepinephrine, enhancing cognitive function and potentially mitigating anxiety. By lowering oxidative stress, frankincense can protect neural tissues and maintain normal physiological function in the brain. This makes frankincense useful for neurodegenerative disorders and stress responses.

9. Ulcerative Colitis and Gastrointestinal Health

Molecular or Cellular Target/Pathway: Inflammatory mediators in the gut.

Mechanisms: Frankincense oil attenuates various inflammatory markers to mitigate symptoms of ulcerative colitis. Specifically, it significantly reduces the levels of proinflammatory cytokines, such as TNF-α and IL-6, in the colon, and inhibits the activation of NF-κB and MAPK signaling pathways. Interestingly, molecular docking analysis identified hepten-2-yl tiglate, 6-methyl-5-; Isoneocembrene A; and para-Cymene as the constituents in the oil that bound to target proteins in the NF-κB and MAPK signaling pathways. Overall, the oil reduces symptoms of colitis, such as diarrhea, blood in stool, and overall inflammation in the gastrointestinal tract.

10. Antileishmanial Activity

Molecular or Cellular Target/Pathway: *Leishmania* protozoa.

Mechanisms: Frankincense shows notable activity against parasites causing leishmaniasis, targeting their growth and metabolism. It may interfere with the lifecycle or replication of the protozoa.

GERANIUM

Geranium essential oil, derived from the leaves and stems of the *Pelargonium graveolens*, is an underutilized versatile powerhouse of therapeutic benefits, often heralded for its remarkable properties in the realm of holistic healing. This oil shows promising anticancer activity, evidenced by its ability to selectively target and reduce the viability of leukemia cell lines through the induction of apoptosis, presenting a potential complementary therapy to traditional cancer treatments. Additionally, geranium oil boasts potent antimicrobial properties, significantly inhibiting the growth of various pathogens, including *E. coli* and *C. albicans*, with its efficacy enhanced in synergy with conventional antibiotics. Its anti-inflammatory effects are equally noteworthy, as geranium oil helps mitigate inflammation, benefiting conditions such as stomatitis and vaginal candidiasis, while also exhibiting robust antioxidant activity that neutralizes free radicals and protects against oxidative damage. Furthermore, geranium essential oil supports hormonal balance, enhances wound healing, and shows promising neuroprotective effects, positioning it as an invaluable asset in both therapeutic and cosmetic applications. With its array of benefits, geranium essential oil can promote greater wellness in various aspects of life.

SUGGESTED THERAPEUTIC USE:

Cancer: Leukemia

Infections: Bacterial infections, fungal infections, antibiotic-resistant infections

Inflammatory Conditions: Stomatitis, vaginal candidiasis, general inflammation

Organ and Tissue Protection: Oxidative stress-related conditions (cancer, IBD, diabetes, neurodegenerative conditions, cardiovascular disease, liver or kidney disease, metabolic syndrome, fibromyalgia, chronic fatigue syndrome, skin aging), diabetes-related complications, pesticide-related neurotoxicity

Metabolic Disorders: Diabetes, hyperglycemia

Hormone Balance and Female Reproductive Concerns: PMS, menopause-related symptoms, hormonal imbalances

Neurological Disorders: Parkinson's disease, neurodegenerative disorders

Insect Repellency: Tick infestations, insect infestations

Wounds: Chronic wounds, ulcers, diabetic foot ulcers, wound infections

Skin Issues: Skin aging, skin health issues

MOLECULAR AND CELLULAR TARGETS, BIOLOGICAL PATHWAYS, AND MECHANISMS:

1. Anticancer Activity

Molecular or Cellular Target/Pathway: Cancer cell signaling pathways, cancer proapoptotic pathways.

Mechanism: Geranium oil has demonstrated the ability to reduce the viability of leukemia cell lines *in vitro*, suggesting selective targeting of malignant cells. The oil likely activates apoptotic pathways, leading to cancer cell death. By selectively targeting cancerous cells, geranium essential oil may enhance existing treatment methods and reduce side effects compared to conventional therapies.

2. Antimicrobial Properties

Molecular or Cellular Target/Pathway: Bacterial and fungal cell membranes.

Mechanism: Geranium oil shows inhibitory effects on various pathogenic bacteria and fungi. The essential oil disrupts cellular membranes, leading to cell death and inhibiting pathogen growth. It enhances the activity of conventional antibiotics when combined with them (e.g., ciprofloxacin, gentamicin), suggesting a synergistic effect that may reduce the required dosages of antibiotics and their potential side effects.

3. Anti-inflammatory Effects

Molecular or Cellular Target/Pathway: Neutrophil accumulation and proinflammatory cytokines.

Mechanism: Geranium oil reduces neutrophil accumulation and other markers of inflammation. Neutrophil accumulation is closely linked to inflammation as these white blood cells are one of the body's first responders to sites of injury or infection. When tissues are damaged or infected, signaling molecules called cytokines are released, which attract neutrophils to the area. Once they arrive, neutrophils help fight off infections by ingesting pathogens, such as bacteria, and releasing enzymes that can break down damaged tissue. While this process is important for healing, an excessive accumulation of neutrophils can lead to chronic inflammation, where the body continues to react as if there is an ongoing threat. This can cause additional tissue damage and contribute to conditions like arthritis, asthma, and other inflammatory diseases. It likely suppresses signals leading to inflammation, improving conditions like stomatitis and vaginal candidiasis.

4. Antioxidant Activity

Molecular or Cellular Target/Pathway: ROS and oxidative stress.

Pathway: Geranium essential oil enhances antioxidant enzyme activity, neutralizing free radicals and minimizing oxidative stress. The oil's antioxidant properties help protect against oxidative damage in various tissues, aiding in the prevention of related complications in conditions such as diabetes and neurotoxicity from certain pesticides. This activity helps preserve cellular integrity and function in the face of oxidative stress.

5. Antidiabetic Properties

Molecular or Cellular Target/Pathway: Pancreatic beta-cell function

Mechanisms: Geranium oil enhances insulin production and protects against oxidative damage related to diabetes. Pancreatic beta-cells monitor blood glucose levels and release insulin when glucose is high, thereby regulating blood sugar. Its components promote insulin secretion in response to glucose, which can mitigate complications associated with the disease. The oil helps regulate blood sugar levels more effectively than some pharmacological agents, representing a potential adjunct in diabetes management.

6. Hormonal Effects

Molecular or Cellular Target/Pathway: Estrogen receptors.

Mechanisms: Inhalation of geranium oil has been shown to increase salivary estrogen levels in women, suggesting an influence on hormonal balance. Constituents within geranium oil—like geraniol and citronellol—may interact with estrogen receptors (ERs), specifically the estrogen receptor alpha (ERα) and estrogen receptor beta (ERβ), which are proteins that mediate the effects of estrogen in different tissues. When compounds in geranium essential oil bind to these receptors, they can activate or inhibit various cellular processes that estrogen normally influences, such as cell growth, differentiation, and apoptosis. Therefore, these phytochemicals mimic estrogen's effects and potentially trigger increased levels of salivary estrogen as a response. Additionally, these interactions may stimulate the endocrine system, leading to altered hormone levels. This suggests potential applications for conditions influenced by estrogens, including PMS and menopause-related issues.

7. Neuroprotective Effects

Molecular or Cellular Target/Pathway: Oxidative stress.

Mechanisms: By improving oxidative balance and preventing neuronal death, particularly hypothalamic neurons, geranium oil has potential in neurodegenerative diseases like Parkinson's. This may involve antioxidants promoting neuronal health and function by reducing neuroinflammation.

8. Repellent Activity

Molecular or Cellular Target/Pathway: Insect nervous systems.

Mechanisms: Compounds like 10-epi-gamma-eudesmol in geranium function as natural repellents against ticks and insects. The oil disrupts physiological functions in insects, leading to reduced population sizes of pests like cattle tick (*Rhipicephalus microplus*).

9. Improved Wound Healing

Molecular or Cellular Target/Pathway: Wound inflammation and fibroblast activity.

Mechanisms: Geranium oil enhances tissue regeneration and reduces inflammation in chronic wounds and ulcers by stimulating fibroblast activity. The oil supports repair processes essential for wound healing, particularly in diabetic foot ulcers and infections. This may involve a combination of anti-inflammatory effects and stimulatory actions on cellular growth processes.

10. Skin Health

Molecular or Cellular Target/Pathway: Antioxidant activity and inflammatory mediators.

Mechanisms: Geranium oil's antioxidant properties are beneficial in cosmetic formulations aimed at anti-aging and improving skin health because it protects the skin from damage caused by free radicals. Its abilities to combat microorganisms and oxidative stress enhance skin condition and may aid in protection against environmental damage. It can help reduce skin irritation and inflammation. Geranium may also act as an astringent, tightening the skin and reducing the appearance of pores.

GERMAN CHAMOMILE

German chamomile essential oil, derived from the flowers of *Matricaria chamomilla*, is a cherished natural remedy known for its diverse therapeutic benefits, particularly in the realms of cancer and inflammation. This exceptional oil exhibits promising anticancer properties by inhibiting the growth of colon carcinoma cells through the disruption of DNA replication and inducing apoptosis in glioma cells, positioning it as a potential adjunct to traditional cancer therapies. Beyond its anticancer capabilities, German chamomile is renowned for its powerful anti-inflammatory effects, where it modulates key inflammatory pathways, reduces the release of proinflammatory mediators, and alleviates conditions like eczema and arthritis. Coupled with robust antioxidant activity that effectively scavenges reactive oxygen species, this essential oil protects cellular integrity and promotes overall wellness. German chamomile also offers neuroprotective benefits, microbial defense against various pathogens, and supports gastrointestinal health, making it a versatile asset in holistic health practices. With these myriad benefits, German chamomile oil is a formidable natural weapon for those seeking natural pathways to health and healing.

SUGGESTED THERAPEUTIC USE:

Cancer: Colon carcinoma, glioma

Inflammatory Conditions: Eczema, arthritis, general inflammation, mast cell-related allergic reactions

Organ and Tissue Protection: Oxidative stress-related conditions (cancer, IBD, diabetes, neurodegenerative conditions, cardiovascular disease, liver or kidney disease, metabolic syndrome, fibromyalgia, chronic fatigue syndrome, skin aging), cellular damage

Neurological Disorders: Migraines, headaches, Parkinson's disease, Lewy body dementia, multiple system atrophy, neurological conditions

Infections: Bacterial infections, viral infections (e.g., HSV-1, HSV-2), skin infections, oral infections

Gastroprotection: Gastric ulcers, gastrointestinal inflammation, digestive issues

Cardiovascular issues: High blood pressure, anxiety related to acute coronary syndrome, cardiovascular health issues

Pain: Chronic inflammatory pain, general pain

Wounds: Skin injuries, postoperative wounds

Parasites: Leishmaniasis, anisakiasis

MOLECULAR AND CELLULAR TARGETS, BIOLOGICAL PATHWAYS, AND MECHANISMS:

1. Anticancer Activity

Molecular or Cellular Target/Pathway: DNA polymerase alpha and beta, cancer cell proapoptotic pathways

Mechanisms: German chamomile inhibits the growth of colon carcinoma cells by disrupting DNA replication, effectively blocking these enzymes. DNA Polymerase Alpha and Beta are two important proteins that help our cells copy their DNA. DNA

Polymerase Alpha is mainly involved in starting the process of DNA replication, making sure that when cells divide, they accurately duplicate their genetic information. DNA Polymerase Beta helps repair damaged DNA, fixing any mistakes that occur when cells are copying their DNA or when DNA gets damaged. When these proteins don't work correctly, it can lead to problems in DNA copying and repair, which may cause cells to grow uncontrollably. This uncontrolled growth can lead to cancer, where cells multiply too much, forming tumors and spreading to other parts of the body. Maintaining healthy DNA Polymerase Alpha and Beta is really important to prevent cancer and make sure our cells stay healthy. Active compounds like alpha-bisabolol trigger apoptotic pathways in glioma cells, promoting cancer cell death. By targeting DNA replication and promoting cell death, German chamomile oil exhibits potential as a natural adjunct in cancer treatment.

2. Anti-inflammatory Effects

Molecular or Cellular Target/Pathway: 5-LOX, TNF-α, mast cells, proinflammatory cytokines (IL-6).

Mechanisms: German chamomile moderates inflammatory responses by moderately inhibiting this enzyme involved in leukotriene synthesis. Its anti-inflammatory properties are enhanced by lowering leukocyte activity and modulating inflammatory pathways, facilitating relief from pain and tissue damage caused by inflammation. The oil decreases TNF-α, IL-6, and other inflammatory markers, leading to reduced inflammation in various conditions, such as eczema and arthritis. It inhibits mast cell degranulation, reducing the release of proinflammatory mediators like histamine and TNF-α. A type of immune cell, mast cells play a key role in the inflammatory response to injury or infection. When mast cells detect something harmful, like an allergen or a pathogen, they undergo a process called degranulation. This means they release tiny packets of chemicals, such as histamine and other substances, into the surrounding area. These chemicals help increase blood flow and attract other immune cells to the site of the problem, causing redness, swelling, and pain—these are all signs of inflammation. While this response is vital for fighting off infections and healing injuries, it can also lead to allergic reactions and other inflammatory diseases if the mast cells react too strongly or inappropriately.

3. Antioxidant Activity

Molecular or Cellular Target/Pathway: ROS, oxidative stress.

Mechanisms: German chamomile demonstrates strong antioxidant capabilities, exceeding those of commonly used antioxidants like vitamins C and E. The essential oil effectively neutralizes free radicals, preventing oxidative DNA damage and offering protective effects for cells.

4. Neuroprotective Properties

Molecular or Cellular Target/Pathway: Neurogenic inflammation, oxidative stress, proinflammatory cytokines (IL-1β, IL-6, and TNF-α), inflammatory mediators (iNOS, COX-2), antiapoptotic pathways, mitochondrial function modulation, Bax, cleaved caspase-3, cleaved caspase-9, and Bcl-2.

Mechanisms: German chamomile may protect neurons from damage and reduce neurogenic inflammation associated with migraines and other neurological conditions. One of its active sesquiterpenes (alpha-Bisabolol) protects against oxidative stress, evidenced by the inhibition of MDA formation and GSH depletion, as well as the improvement in antioxidant enzymes, SOD and CAT. Alpha-bisabolol prevents overactivation of glial cells, as well as the induction and release of proinflammatory cytokines (IL-1β, IL-6, and TNF-α) and inflammatory mediators (iNOS and COX-2) in the striatum. The striatum is a part of the brain that helps control movement, learning, and reward. Additionally, alpha-bisabolol decreases apoptosis of dopaminergic neurons by attenuating the downregulation of antiapoptotic protein Bcl-2 and upregulation of proapoptotic proteins Bax, cleaved caspases 3 and 9. In Parkinson's disease, too many dopaminergic neurons undergo apoptosis. Balancing these proteins controls the self-destruction (apoptosis) process. It tones down the proteins that tell the cell to die (Bax, cleaved caspase-3, and cleaved caspase-9) and boosts the proteins that keep the cell alive (Bcl-2). By doing this, alpha-bisabolol reduces the excessive cell death of dopaminergic neurons seen in Parkinson's disease. Lastly, alpha-bisabolol preserves mitochondrial function by inhibiting mitochondrial lipid peroxidation, cytochrome c release, and reinstating the

levels/activity of ATP and MC-I. By preventing the release of cytochrome—a substance that should stay inside the mitochondria to help with energy production—alpha-bisabolol helps restore healthy levels and activity of important molecules needed for energy (ATP and MC-I). In the end, the mitochondria can function properly and provide the cell with the energy it needs to survive.

5. Antimicrobial Activity

Molecular or Cellular Target/Pathway: Bacterial cell walls, viral adsorption, direct cytotoxicity.

Mechanisms: German chamomile shows effectiveness against various bacteria and viruses (e.g., HSV-1, HSV-2). It disrupts bacterial membranes, leading to cell lysis and death. For viruses like HSV, it prevents attachment to host cells, thereby blocking infection. German chamomile oil also exerts toxicity against specific cancer and pathogenic cells, contributing to improved health and preventing infections.

6. Gastroprotective Effects

Molecular or Cellular Target/Pathway: *H. pylori* growth, sulfhydryl groups, gastrointestinal inflammation, leukocyte activity.

Mechanisms: Alpha-bisabolol, a major constituent, protects against gastrointestinal ulcers induced by NSAIDs and also reduces gastrointestinal inflammation. It decreases leukocyte activity and mitigates inflammation, promoting healing of gastrointestinal tissues. Alpha-bisabolol increases the availability of certain protective substances (sulfhydryl groups like cysteine and glutathione) in the stomach, which reduce the damage caused by harmful substances like ethanol and indomethacin, leading to oxidative gastric injury. Essentially, alpha-bisabolol strengthens the stomach's natural defenses against damage. By preserving mucosal integrity and reducing inflammatory responses, chamomile oil supports digestive health and reduces ulcer risks.

7. Muscle Relaxant and Hypotensive Actions

Molecular or Cellular Target/Pathway: Angiotensin-converting enzyme (ACE), endogenous antioxidant activity (SOD, GSH).

Mechanisms: The essential oil exhibits relaxant effects on blood vessels, thereby reducing blood pressure. It does so by inhibiting the activity of angiotensin-converting enzyme (ACE). Inhibiting ACE helps to lower blood pressure by preventing the production of a substance (angiotensin I) that narrows blood vessels. This allows blood vessels to relax and widen, making it easier for blood to flow through, which in turn reduces blood pressure. Additionally, chamomile restores GSH in tissues and increased SOD levels and activity after depletion. Restoring glutathione (GSH) and superoxide dismutase (SOD) activity relates to blood pressure by mitigating oxidative stress, which plays a significant role in the development and progression of hypertension. By inducing vasodilation, it allows for improved blood flow, which can provide cardioprotective effects and enhance overall cardiovascular health.

8. Analgesic Properties

Molecular or Cellular Target/Pathway: Inflammatory mediators (leukocyte activity, neutrophil degranulation, TNF-α, IL-6), neurotransmitters.

Mechanisms: German chamomile reduces hyperalgesia and general pain sensitivity. It modulates the release of inflammatory mediators and decreases the activation of pain pathways. During injury or infection, cells (e.g., neutrophils and leukocytes) are sent to the area to promote healing. However, these chemicals can also irritate your nerves, making them more sensitive, which is why you feel pain. If not balanced, these cells can contribute to tissue damage, exacerbating pain. By reducing inflammatory responses and modulating neurotransmitters, it aids in alleviating migraine symptoms, making it a potential natural remedy for headache relief.

9. Wound Healing

Molecular or Cellular Target/Pathway: Fibroblast cell viability.

Mechanisms: German chamomile oil accelerates the wound-healing process, promoting fibroblast cell viability, migration, and proliferation. Fibroblasts are crucial in wound healing because they are the cells that produce collagen and other components of the extracellular matrix, which are essential for tissue repair. Its healing properties exceeded those of corticosteroids by stimulating faster re-epithelialization, and possibly by enhancing angiogenesis and reducing oxidative damage to wound tissues. It aids in managing wound inflammation and enhances tissue regeneration processes.

10. Antiparasitic Effects

Molecular or Cellular Target/Pathway: Direct toxicity to parasites.

Mechanisms: The oil is toxic to certain parasites, effectively aiding in their eradication. German chamomile exhibits activity against parasites such as *L. infantum* and *Anisakis*. *L. infantum* causes a disease known as leishmaniasis, specifically the form called visceral leishmaniasis (VL). This severe form of leishmaniasis affects internal organs such as the liver, spleen, and bone marrow. Anisakis is a genus of parasitic worms that can infect fish and marine mammals. These parasites can cause a condition known as anisakiasis in humans, which occurs when people eat raw or undercooked fish that contains these larvae. Symptoms of anisakiasis may include abdominal pain, nausea, vomiting, and diarrhea, as the larvae migrate into the stomach or intestines.

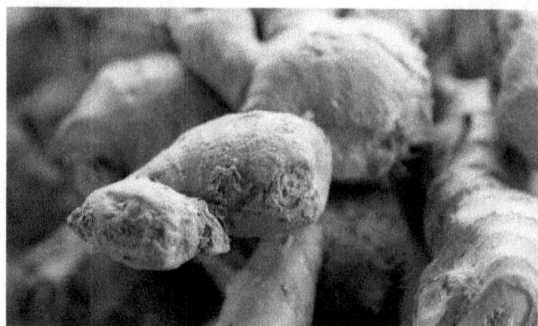

GINGER

Ginger essential oil, derived from the rhizome of *Zingiber officinale*, has been employed for centuries in traditional medicine, and contemporary research is increasingly validating its diverse therapeutic applications. Studies have demonstrated its efficacy in alleviating pain, reducing inflammation, combating microbial infections, and even influencing complex conditions like cardiovascular disease and neurological disorders. From reducing low back pain and knee osteoarthritis to inhibiting viral replication and modulating the gut microbiome, ginger oil's multifaceted benefits are attributed to its rich composition of bioactive compounds, including alpha-zingiberene beta-sesquiphellandrene and various other terpenes. These components interact with numerous molecular targets and biological pathways, making ginger essential oil a comprehensive approach to health and wellness.

SUGGESTED THERAPEUTIC USE:

Cancer: Colorectal, ovarian, gastric, liver, breast, lung

Inflammatory Conditions: Osteoarthritis, rheumatoid arthritis, chronic inflammatory conditions (IBD, psoriasis, COPD, asthma, lupus, eczema, gout, chronic sinusitis, fibromyalgia, chronic fatigue syndrome, periodontitis)

Pain: Low back pain, dysmenorrhea, muscle pain, tendon/ligament pain

Digestive Disorders: Nausea, vomiting, motion sickness, stomach ulcers, gastroesophageal reflux disease (GERD), constipation, ulcerative colitis

Respiratory Conditions: Chronic respiratory conditions involving bronchoconstriction (asthma, COPD, allergic bronchopulmonary aspergillosis, exercise-induced bronchoconstriction, bronchiectasis, reactive airway disease), respiratory infections, colds, flu

Cardiovascular Conditions: Atherosclerosis, high cholesterol and triglycerides

Metabolic Disorders: Type 2 diabetes, nonalcoholic fatty liver disease (NAFLD), nonalcoholic steatohepatitis (NASH)

Infections: Viral (HSV-1, HSV-2, human alphaherpesvirus 2), bacterial infections, fungal infections, mycobacterial infections

Neurological Conditions: Neurodegenerative diseases (Alzheimer's disease, Parkinson's disease, multiple sclerosis), epilepsy, migraines

Mental Health Issues: Anxiety, depression

Oral Health: Gingivitis, dental biofilms

Skin Conditions: Acne, skin aging, hyperpigmentation

MOLECULAR AND CELLULAR TARGETS, BIOLOGICAL PATHWAYS, AND MECHANISMS:

1. Analgesic and Anti-inflammatory Properties

Molecular or Cellular Target/Pathway: COX-2, LOX, proinflammatory cytokines (IL-6, IL-10, and TNF-α), NF-κB signaling.

Mechanisms: Ginger oil reduces inflammation by inhibiting the production of prostaglandins (COX-2) and leukotrienes (LOX), key mediators of pain and inflammation. It also interferes with the NF-κB pathway, a central regulator of inflammatory gene expression, and reduces the release of proinflammatory cytokines (IL-6, IL-10, and TNF-α), thereby mitigating inflammatory responses in conditions such as rheumatoid arthritis and osteoarthritis.

2. Antimicrobial and Antiviral Activity

Molecular or Cellular Target/Pathway: Bacterial cell membranes, viral replication, biofilm formation.

Mechanisms: Ginger oil's antimicrobial effects stem from its ability to disrupt bacterial cell membranes, leading to cell lysis. Its antiviral properties involve inhibiting viral replication at various stages. It also inhibits the formation of biofilms by various microbes. This activity is demonstrated against a range of pathogens, including *S. aureus*, *E. coli*, and herpes simplex virus.

3. Gastrointestinal Health Modulation

Molecular or Cellular Target/Pathway: 5-HT3 receptor channel system, gastrointestinal motility, *H. pylori*.

Mechanisms: Ginger oil can reduce nausea and vomiting by inhibiting the 5-HT3 receptor, a key player in the emetic reflex. 5-HT3 receptors are a type of serotonin receptor found in the brain and the gut. When certain triggers, such as the sight or smell of food, motion sickness, or medications like chemotherapy, activate these receptors, they send signals to the brain that can lead to feelings of nausea and the urge to vomit. The vagus nerve, which connects the brain to the digestive system, plays a crucial role in this process. It communicates these signals from the gut to the brain, enhancing the sensation of nausea. Essentially, when the 5-HT3 receptors are activated, they contribute to a chain reaction involving the vagus nerve that can result in vomiting. It also enhances gastrointestinal motility, aiding in the movement of food through the digestive tract and relieving conditions like gastroesophageal reflux. Furthermore, it exhibits potent antibacterial activity against *H. pylori*, a bacterium associated with stomach ulcers and cancer.

4. Cardiovascular and Metabolic Effects

Molecular or Cellular Target/Pathway: Trimethylamine-N-oxide (TMAO) levels, lipoprotein metabolism, metabolism, gut microbiome.

Mechanisms: Ginger oil's cardiovascular benefits are linked to its ability to reduce TMAO, a metabolite associated with atherosclerosis. TMAO is a compound that forms when gut bacteria break down certain nutrients found in foods like red meat, eggs, processed meats, fish, seafood, and dairy products. High levels of TMAO in the blood have been linked to an increased risk of cardiovascular disease, particularly atherosclerosis, which is the buildup of fatty deposits in the arteries. TMAO can promote the formation of cholesterol plaques by influencing how cholesterol is processed in the body and encouraging inflammation in the blood vessels. As these plaques grow, they can narrow arteries and restrict blood flow, leading to heart problems like heart attacks or strokes. In short, TMAO acts like a signal that can worsen heart health by contributing to the hardening and narrowing of the arteries. It also improves lipid metabolism, reduces blood sugar, and enhances insulin sensitivity. Additionally, it positively modulates the gut microbiome, increasing the abundance of beneficial bacteria and reducing harmful ones, contributing to overall metabolic health.

5. Neurological Protection

Molecular or Cellular Target/Pathway: mRNA expression of GABA receptors, neuroinflammation, nerve remyelination, AChE and adenosine deaminase (ADA) activities

Mechanisms: Ginger oil exhibits neuroprotective effects by modulating neurotransmitter receptors, reducing neuroinflammation, and promoting remyelination. Myelin is a protective coating that surrounds the nerve fibers in our brain and throughout our body, much like the insulation on electrical wires. This myelin sheath helps speed up the transmission of electrical signals between nerve cells, allowing them to communicate effectively. Remyelination is the process by which the body repairs or replaces damaged myelin, helping restore normal nerve function. This is important in conditions like multiple sclerosis, Guillain-Barré syndrome, and peripheral neuropathy, where myelin is damaged, leading to disruptions in communication between the brain and the rest of the body. It can also inhibit enzymes like AChE and ADA, which are implicated in neurodegenerative diseases. ADA is an enzyme in the body that breaks down adenosine, a molecule that plays a crucial role in various biological processes, including regulating nerve function. When this enzyme is inhibited or blocked, it leads to increased levels of adenosine. Higher adenosine levels can help protect nerves from damage and reduce inflammation, which is beneficial in conditions like stroke or neurodegenerative diseases. Essentially, by stopping the breakdown of adenosine, we enhance its natural protective effects on the brain and nervous system, helping to maintain healthier nerve cells and improve their survival during stress or injury. These mechanisms suggest ginger provides potential benefits in conditions like epilepsy, multiple sclerosis, Alzheimer's disease, and various other neurological disorders.

6. Skin Health and Anti-Aging

Molecular or Cellular Target/Pathway: Melanin production, tyrosinase, melanogenesis-related proteins, TNF-α and MMP-1 expression, endogenous antioxidant activities (GSH, SOD, and CAT)

Mechanisms: Ginger oil can lighten the skin by inhibiting melanin production and tyrosinase activity. Its antioxidant properties, including enhancing glutathione (GSH), superoxide dismutase (SOD), and catalase (CAT) activities, protect against oxidative stress and delay skin aging. It also reduces inflammation and the breakdown of collagen by downregulating TNF-α and MMP-1, respectively.

GRAPEFRUIT

Grapefruit essential oil, extracted from the vibrant peel of the grapefruit (*Citrus × paradisi*), emerges as a multifaceted natural remedy celebrated for its impressive array of health benefits. Notably, this essential oil exhibits potent anticancer properties, particularly in inducing apoptosis in leukemia cells, suggesting its potential as an adjunct therapy in cancer management by mitigating cell proliferation. Grapefruit oil is also known for its ability to support weight management and lipid metabolism by inhibiting the formation of fat-storing cells while promoting fat breakdown, which can assist with appetite control and energy mobilization. Its anti-inflammatory and antimicrobial effects further enhance its therapeutic profile, as grapefruit oil reduces proinflammatory cytokines and combats various bacterial and fungal species. Additionally, with antioxidant properties that protect against oxidative stress and benefits for gut health and hormonal regulation, grapefruit essential oil stands out as a versatile ally in promoting overall well-being. Its mood-enhancing and pain-relieving effects, combined with insecticidal properties, illustrate the diverse applications of grapefruit oil in both health and lifestyle, advocating for its place in natural wellness routines.

<u>SUGGESTED THERAPEUTIC USE</u>:

Cancer: Leukemia

Weight Management and Lipid Disorders: Obesity, weight gain, high cholesterol

Inflammatory Conditions: Arthritis, skin inflammation, chronic inflammatory conditions (rheumatoid arthritis, IBD, psoriasis, COPD, asthma, lupus, eczema, gout, chronic sinusitis, fibromyalgia, chronic fatigue syndrome, periodontitis), conditions with excessive leukotriene production (asthma, seasonal allergies, COPD, eosinophilic esophagitis, psoriasis, allergic conjunctivitis, lupus, cystic fibrosis, gout)

Infections: Bacterial infections (e.g., *S. aureus*, *E. coli*, *P. aeruginosa*), fungal infections (e.g., *C. albicans*), oral infections

Organ and Tissue Protection: Oxidative stress-related conditions (cancer, IBD, diabetes, neurodegenerative conditions, cardiovascular disease, liver or kidney disease, metabolic syndrome, fibromyalgia, chronic fatigue syndrome, skin aging), cellular damage, aging

Gastroprotection: *H. pylori* infection, gastrointestinal disorders, ulcers

Hormone Balance and Detoxification: Hormonal imbalances, zearalenone exposure

Pain: Abdominal pain during procedures

Mood: Anxiety

Metabolic Disorders: Diabetes, hyperglycemia

Insect Control and Repellency: Insect infestations

<u>**MOLECULAR AND CELLULAR TARGETS, BIOLOGICAL PATHWAYS, AND MECHANISMS**</u>:

1. Anticancer Properties

Molecular or Cellular Target/Pathway: Cancer cell proapoptotic pathways

Mechanisms: Grapefruit oil induces apoptosis in leukemia cancer cells by interfering with critical cellular processes. The activation of pathways leading to programmed cell death suggests that grapefruit oil can help mitigate cancer cell proliferation.

2. Weight Management and Lipid Metabolism

Molecular or Cellular Target/Pathway: Adipogenesis, lipolytic pathway.

Mechanisms: Inhalation of grapefruit oil reduces appetite and stimulates energy mobilization from fat stores, potentially aiding in weight management. Grapefruit oil inhibits adipogenesis, which is the process by which preadipocytes turn into adipocytes (fat-storing cells). Adipogenesis is the process where your body creates new fat cells. When adipogenesis is inhibited, your body makes fewer new fat cells. Therefore, if your body is not creating new fat cells, it is harder for your body to store new fat. This can help prevent the accumulation of extra fat. The oil also promotes lipolysis (the breakdown of fats for energy), indicated by increased plasma glycerol levels, which suggests mobilization of stored fats.

3. Anti-inflammatory Effects

Molecular or Cellular Target/Pathway: Proinflammatory cytokines (IL-6, TNF-α), COX-2, NF-κB, oxidative stress.

Mechanisms: Grapefruit oil offers protective effects against inflammation by reducing free radical production and modulating inflammatory pathways in conditions like arthritis and skin inflammation. Grapefruit essential oil inhibits the production of inflammatory cytokines such as IL-6 and TNF-α. The inhibition of COX-2 suggests that grapefruit oil may dampen inflammatory responses associated with tissue damage and chronic inflammation.

4. Antimicrobial Activity

Molecular or Cellular Target/Pathway: Bacterial quorum sensing, pathogen cellular integrity.

Mechanisms: Grapefruit oil inhibits quorum sensing, a mechanism bacteria use for communication, suggesting it could reduce bacterial virulence. The oil's constituents can disrupt cellular integrity in pathogens, effectively reducing infection rates and supporting oral health.

5. Antioxidant Activity

Molecular or Cellular Target/Pathway: ROS.

Mechanisms: Grapefruit essential oil exhibits antioxidant properties, important for protecting cells from oxidative stress. The oil can scavenge free radicals, which helps mitigate oxidative damage that contributes to aging and various diseases.

6. Gastroprotective Effects

Molecular or Cellular Target/Pathway: Gut microbiome, *H. pylori*.

Mechanisms: By reducing the load of pathogenic bacteria in the gastrointestinal tract, grapefruit oil may support gastrointestinal health and healing. Grapefruit oil has been shown to inhibit the growth of *H. pylori*, a bacterium associated with ulcers and other GI disorders. Limonene positively modulates the gut microbiome, increasing alpha diversity. No significant changes were observed in beta diversity. Alpha and beta diversity are two important ways to measure the variety of microorganisms in the gut microbiome, which is the collection of bacteria and other microbes living in our intestines. Alpha diversity refers to the diversity of species within a specific environment, like a single person's gut. It shows how many different types of bacteria are present and how evenly they are distributed. A high alpha diversity usually indicates a healthy gut microbiome. On the other hand, beta diversity compares the differences in microbial communities between different individuals or environments. It reveals how similar or different the gut microbiomes of two people are. For instance, if two people have very different types of bacteria in their guts, they have high beta diversity. Together, alpha and beta diversity give us a fuller picture of gut health and how it may vary from person to person, potentially impacting overall health and disease susceptibility.

7. Hormonal Regulation and Detoxification

Molecular or Cellular Target/Pathway: Fusarium genes involved in the zearalenone biosynthetic pathway.

Mechanisms: Grapefruit oil helps reduce the production of zearalenone, an endocrine-disrupting mycotoxin produced by *Fusarium* species fungi, thereby decreasing the potential for hormonal imbalances. Zearalenone attaches to estrogen receptors (ERalpha, ERbeta) in a similar manner to 17beta-estradiol (naturally produced estrogen) and both it and its metabolites (beta-zearalenol, alpha-zearalenol) alter hormone production (progesterone, estradiol, testosterone, and cortisol) and balance. This action supports detoxification processes and may contribute to hormonal balance.

8. Pain Relief and Mood Enhancement

Molecular or Cellular Target/Pathway: Nervous system activity.

Mechanisms: The olfactory properties of grapefruit oil may influence the central nervous system, providing both psychological and physiological relief. Research indicated that inhalation of grapefruit oil can alleviate anxiety and reduce sensations of abdominal pain during invasive procedures.

9. Lipid Metabolism and Blood Sugar Management

Molecular or Cellular Target/Pathway: Glycolytic enzymes (alpha-glucosidase and alpha-amylase).

Mechanisms: Grapefruit oil reduces the activity of alpha-glucosidase and alpha-amylase, indicating potential blood sugar-lowering effects. The modulation of these enzymes suggests grapefruit oil may help manage blood sugar levels and could be beneficial in diabetes management.

10. Insecticidal and Repellent Properties

Molecular or Cellular Target/Pathway: Insect nervous systems.

Mechanisms: Grapefruit oil and its constituents demonstrate toxic effects on specific insect species, indicating potential use in pest control. By affecting the nervous system of insects, it serves as both a repellent and a lethal agent against pests.

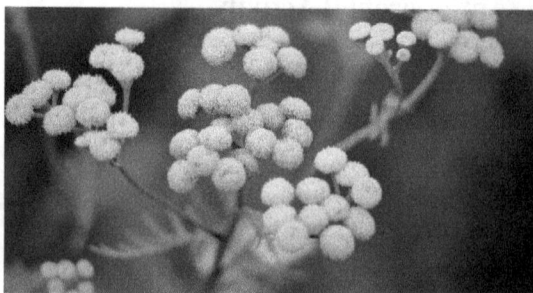

HELICHRYSUM

Helichrysum essential oil, derived from the resilient flowers of *Helichrysum italicum*, is celebrated for its remarkable therapeutic properties that make it a powerhouse in both skincare and health applications. Three primary chemotypes are available commercially: 1) Corsican, rich in neryl acetate and variable concentration of italidiones; 2) Balkan, with low neryl acetate and higher levels of italidiones and pinenes; and 3) Italian, with inconsistent composition that can sometimes be rich in neryl acetate or italicene, but is often poor in italidiones. Renowned for its potent antimicrobial activity, this essential oil effectively combats a diverse range of bacteria and fungi by disrupting their cell membranes while also inhibiting harmful biofilm formation and bacterial communication. Coupled with its compelling anti-inflammatory effects, which inhibit the production of proinflammatory cytokines and scavenge free radicals, helichrysum oil plays a vital role in reducing chronic inflammation and supporting wound healing. Furthermore, its unique constituents exhibit promising anti-HIV and anticancer properties, showcasing the oil's potential in complementary therapies. The oil is also rich in antioxidant compounds that protect against oxidative stress and aging, alongside mood-enhancing benefits that alleviate mental exhaustion. With insecticidal properties and significant advantages for skin health—promoting hydration and barrier function—helichrysum essential oil stands out as a versatile and invaluable addition to natural health and skincare regimens.

SUGGESTED THERAPEUTIC USE:

Microbial Infection: Bacterial infections, fungal infections, biofilm-related infections (chronic wound infections, catheter-associated UTIs, heart valve infections, chronic sinusitis, cystic fibrosis, dental plaque, periodontal disease, osteomyelitis, conjunctivitis, implant-associated infections, ventilator-associated pneumonia, dialysis-associated infections)

Inflammatory Conditions: Chronic inflammatory conditions (rheumatoid arthritis, IBD, psoriasis, COPD, asthma, lupus, eczema, gout, chronic sinusitis, fibromyalgia, chronic fatigue syndrome, periodontitis), conditions with excessive leukotriene production (asthma, seasonal allergies, COPD, eosinophilic esophagitis, psoriasis, allergic conjunctivitis, lupus, cystic fibrosis, gout), wound healing, tissue damage

Wounds: Wounds, diabetic wounds, impaired wound healing

Viral Infections: HIV-1 infection

Cancer: Ovarian cancer, estrogen-dependent cancers

Organ and Tissue Protection: Oxidative stress-related conditions (cancer, IBD, diabetes, neurodegenerative conditions, cardiovascular disease, liver or kidney disease, metabolic syndrome, fibromyalgia, chronic fatigue syndrome, skin aging), aging, cellular damage

Mood and Cognitive Issues: Mental exhaustion, stress

Insect Control: Insect infestations

Skin Issues: Dry skin, aging skin, compromised skin barrier function, environmental skin damage

MOLECULAR AND CELLULAR TARGETS, BIOLOGICAL PATHWAYS, AND MECHANISMS:

1. Antimicrobial Activity

Molecular or Cellular Target/Pathway: Bacterial and fungal cell structures, quorum sensing inhibition, biofilm formation.

Mechanisms: Helichrysum essential oil demonstrates activity against several bacteria and fungi by interacting with their cell membranes. The oil disrupts bacterial communication mechanisms, reducing their virulence. Helichrysum essential oil inhibits biofilm formation of bacteria, enhancing its effectiveness against infections. The oil utilizes various active compounds that may function synergistically, contributing to its strong antimicrobial properties.

2. Anti-inflammatory Effects

Molecular or Cellular Target/Pathway: Proinflammatory cytokines (IL-6 and TNF-α), 5-LOX, free radical scavenging.

Mechanisms: Constituents of helichrysum inhibit the production of inflammatory cytokines, such as IL-6 and TNF-α, disrupting cytokine signal transduction. The oil enhances its protective effects against oxidative stress, which is exacerbated during inflammation. Free radicals can lead to inflammation, which is the body's way of responding to injury or stress but can become problematic when it goes on too long. By neutralizing these free radicals, antioxidants reduce the stress on our cells and tissues, thereby lowering inflammation. Think of antioxidants as little defenders that help keep our body in balance, reducing harmful reactions that could lead to chronic inflammation and related health issues. It also moderately inhibits 5-LOX, which is involved in leukotriene synthesis and inflammation. The combination of these actions helps combat chronic inflammation, supporting wound healing and reducing tissue damage.

3. Wound Healing Properties

Molecular or Cellular Target/Pathway: Tissue remodeling proteins (collagenase, elastase), fibroblast recruitment and activity.

Mechanisms: Helichrysum inhibits collagenase and elastase activities, enzymes that break down collagen and elastin, respectively. This would keep your skin's support system strong and intact, which translates to healthier, younger-looking skin. Its ability to promote collagen production and fibroblast function aids in wound contraction and healing, particularly in diabetic models where wound healing is impaired.

4. Anticancer Activity

Molecular or Cellular Target/Pathway: Cancer cell proliferation and signaling pathways.

Mechanisms: The oil's constituents exhibit cytotoxic properties against various cancer cell lines, highlighting its potential as an adjunct in cancer therapy. Helichrysum essential oil has shown efficacy against ovarian cancer cells and may impact pathways involved in cell proliferation and apoptosis. It is notably antiestrogenic and interacts with genes involved in lymphocyte proliferation, impacting immune responses. By inhibiting the action of estrogen, helichrysum could be highly useful for estrogen-dependent cancers. By blocking estrogen signaling pathways, this activity could lead to reduced proliferation of estrogen-dependent cancer cells, potentially slowing down cancer growth or progression. By interacting with genes involved in lymphocyte proliferation, the substance could enhance or alter the immune responses and cancer cell signaling. A stronger immune response might help the body better recognize and attack cancer cells. Conversely, if lymphocyte proliferation is negatively impacted, it could lead to an impaired immune response, making it more difficult for the body to combat tumor cells. If the immune response is altered, it may also affect the tumor microenvironment.

5. Antioxidant Properties

Molecular or Cellular Target/Pathway: ROS.

Mechanisms: Helichrysum essential oil scavenges free radicals, reducing oxidative stress related to aging and various diseases. By neutralizing free radicals, it helps protect cellular structures, thereby supporting overall health and potentially mitigating signs of skin aging.

6. Mood and Cognitive Effects

Molecular or Cellular Target/Pathway: Olfactory pathway, potential hormone and neurotransmitter modulation.

Mechanisms: Inhalation of helichrysum essential oil (as part of a blend with peppermint and basil) has been implicated in alleviating mental exhaustion and stress, potentially due to its aromatherapeutic properties that influence mood and stress response. When inhaled, essential oils stimulate the olfactory system, which is directly connected to the brain regions involved in emotion and memory, such as the limbic system. This stimulation can help evoke positive emotional responses and reduce feelings of stress or burnout. Additionally, the oils could be influencing the levels of neurotransmitters associated with mood regulation and lowering cortisol levels, thereby reducing the physiological effects of stress and mental fatigue.

7. Insecticidal Properties

Molecular or Cellular Target/Pathway: Insect nervous systems.

Mechanisms: Helichrysum essential oil displays insecticidal activity against specific species, aiding in pest control. Its toxicity to insects may be mediated through nerve disruption or other physiological means.

8. Skin Care Benefits

Molecular or Cellular Target/Pathway: Epidermal differentiation and skin barrier formation, gene expression related to skin health, lipid and ceramide metabolism.

Mechanisms: Application promotes skin hydration and protection against environmental stressors while fostering cellular regeneration. Helichrysum essential oil supports the skin's physiological functions by modulating the gene expression associated with skin health. The skin barrier, primarily composed of the stratum corneum (the outermost layer of the skin), plays a crucial role in maintaining overall skin health and function. It serves as a protective barrier against environmental threats (pathogens, allergens, and chemicals), maintains skin hydration, regulates permeability (allowing some substances to pass while blocking harmful invaders), acts as the immune system's first line of defense, and contains nerve endings important to sense touch, temperature, and pain. The oil aids in maintaining skin structure and barrier function, contributing to a youthful appearance. Lipid and ceramide production are fundamental to skin health and the integrity of the skin barrier, as these components serve as crucial building blocks that keep the stratum corneum hydrated and resilient. Ceramides, a type of lipid (fat), help form the lipid matrix that fills the spaces between skin cells, providing structural support and preventing transepidermal water loss. This lipid composition not only maintains skin hydration but also enhances the barrier's ability to protect against environmental aggressors and pathogens. A deficiency in lipids and ceramides can lead to compromised barrier function, resulting in symptoms such as dryness, irritation, and increased susceptibility to skin conditions like eczema and psoriasis. Therefore, maintaining optimal levels of these lipids is essential for preserving skin health and ensuring effective barrier function.

LAVENDER

Lavender (*Lavandula angustifolia*) essential oil, affectionately known as "the Swiss army knife of essential oils," is celebrated for its remarkable versatility and healing properties that benefit both mind and body. This extraordinary oil exhibits significant anticancer potential, showing the ability to induce apoptosis in various cancer cell lines while leaving healthy cells unharmed, thanks to key constituents like linalool and linalyl acetate. Beyond its anticancer capabilities, lavender oil also stands out for its anti-inflammatory effects, effectively reducing proinflammatory cytokines and promoting antioxidant enzyme activities to combat chronic inflammation. With anxiolytic and neuroprotective properties, lavender oil can ease anxiety, enhance mood, and improve sleep quality through its influence on neurotransmitter levels. Its antimicrobial action helps eliminate a wide range of pathogens, making it invaluable for infection control, while its wound-healing properties support tissue repair and skin health. Additionally, lavender oil plays a role in hormonal balance, pain relief, gastrointestinal health, and even maintains skin vitality through its antioxidant effects. This multifaceted essential oil truly embodies a holistic approach, making it an essential companion in natural wellness.

SUGGESTED THERAPEUTIC USE:

Cancer: Breast cancer, prostate cancer, colorectal cancer, acute lymphoblastic leukemia

Inflammatory Conditions: General inflammation, chronic inflammatory conditions (rheumatoid arthritis, IBD, psoriasis, COPD, asthma, lupus, eczema, gout, chronic sinusitis, fibromyalgia, chronic fatigue syndrome, periodontitis)

Neurological Issues: Anxiety, stress, cognitive decline

Infections: Bacterial infections (gram-positive and gram-negative), fungal infections, antibiotic-resistant infections

Wounds: Wounds, post-surgical wounds, irritated skin, slow-healing wounds

Sleep Disorders: Insomnia, anxiety-related sleep disorders, stress-related sleep disturbances

Pain: Acute pain, chronic pain

Female Concerns: Menopausal symptoms

Metabolic Disorders: High blood pressure, high cholesterol, impaired glucose metabolism, metabolic syndrome, obesity

Intestinal Disturbances: Gut dysbiosis, imbalanced gut microbiota

Respiratory Conditions: Asthma, bronchitis, COPD, respiratory infections, coughs, nasal congestion

Renal/Urinary Conditions: Chronic kidney disease

Cardiovascular Disorders: Heart attack recovery

Reproductive Disorders: Male infertility, hormone balance

Skin Protection: Skin aging, oxidative skin damage, general skin health issues

MOLECULAR AND CELLULAR TARGETS, BIOLOGICAL PATHWAYS, AND MECHANISMS:

1. Anticancer Properties

Molecular or Cellular Target/Pathway: Cancer cell proapoptotic pathways.

Mechanisms: Lavender oil induces apoptosis in cancer cell lines such as breast, prostate (particularly PC-3 cells), colorectal, and acute lymphoblastic leukemia. Constituents like linalool and linalyl acetate in lavender oil exhibit cytotoxicity against cancer cells and prevent metastasis. The oil appears to activate apoptotic pathways, promoting programmed cell death while sparing healthy cells.

2. Anti-inflammatory Effects

Molecular or Cellular Target/Pathway: Proinflammatory cytokines (IL-1β, IL-6, and TNF-α), cytokine signaling, oxidation.

Mechanisms: Lavender oil exhibits the ability to reduce levels of inflammatory mediators, such as IL-1β, IL-6, and TNF-α. Inhibition of lipoxygenase (LOX) and disruption of NF-κB signaling pathways help reduce inflammation. Lavender enhances antioxidant enzyme activities like SOD and inhibits the release of neurotransmitters that contribute to inflammatory responses.

3. Anxiolytic and Neuroprotective Effects

Molecular or Cellular Target/Pathway: Neurotransmitters (GABA, serotonin, glutamate, NMDA, dopamine), 5-HT1A receptors, voltage-gated calcium channels (VGCCs), AChE.

Mechanisms: Inhalation of lavender leads to improved parasympathetic nervous system activity, decreased heart rate, blood pressure, and stress markers like cortisol, thus promoting relaxation and mental comfort. Ingestion of lavender may interact with different molecular targets and biological pathways. The oil reduces neuronal excitability by inhibiting VGCCs. Lavender oil modulates levels of serotonin (binds to the serotonin transporter and affects the activity of 5-HT1A receptors), GABA (potentiates GABA-A receptor activity), glutamate (linalool inhibits glutamate binding in the brain), and NMDA (antagonizes receptors), potentially enhancing mood and reducing anxiety. The 5-HT1A receptor plays a significant role in the regulation of anxiety. These receptors act as both "autoreceptors" (regulating the release of serotonin from the neuron that produces it) and "heteroreceptors" (affecting the activity of other neurons). Autoreceptors, when activated, decrease the release of serotonin. Heteroreceptors, when activated, have an inhibitory effect on the postsynaptic neuron. In a complex way, they help to modulate and control the activity of serotonergic pathways. NMDA (N-methyl-D-aspartate) receptors are a type of glutamate receptor that plays a crucial role in synaptic plasticity, which is essential for learning and memory. Glutamate is the primary excitatory neurotransmitter in the brain and NMDA receptors are activated by glutamate. However, their activity is also deeply intertwined with mood regulation and anxiety. NMDA receptors are vital for long-term potentiation (LTP) and long-term depression (LTD), processes that strengthen or weaken synaptic connections. In some cases, NMDA receptor antagonists—like lavender oil—can exhibit anxiolytic (anti-anxiety) effects. NMDA receptor antagonists also display rapid antidepressant effects by temporarily increasing the release of glutamate, paradoxically leading to enhanced synaptic plasticity and stimulation of BDNF. This supports neuronal growth and survival and modulates neural circuits that regulate mood. The oil also interacts with DRD2 and DRD3 dopamine receptors, which may influence behavior and motivation. Lavender reduces physiological arousal, evidenced by reduced blood pressure and heart rate. It also interacts with the limbic system and autonomic nervous system to promote emotional regulation and stress resilience. Lastly, lavender oil inhibits AChE, promoting enhanced cognition.

4. Antimicrobial Activity

Molecular or Cellular Target/Pathway: Pathogen cell structures and wall/membrane integrity disruption.

Mechanisms: The compounds in lavender enhance antimicrobial activity and reduce antibiotic resistance, as demonstrated in studies involving multi-drug-resistant bacteria. Lavender oil displays activity against various pathogens, including gram-positive and negative bacteria, and fungi. Lavender acts by compromising the cell membrane integrity of microbes, leading to their death.

5. Wound Healing Properties

Molecular or Cellular Target/Pathway: Protein synthesis enzymes, collagen production, fibroblast activity, and VEGF.

Mechanisms: Lavender oil stimulates the expression of type I and II collagen in fibroblasts, promoting tissue repair. Type I and type II collagen play crucial roles in wound healing, primarily through their functions in fibroblasts, which are the cells responsible for producing these collagens. Type I collagen is the most abundant in the body and provides strength and structure to healing tissue, acting like a scaffolding that supports the wound as it heals. Type II collagen, although less abundant in the skin, helps with the flexibility and elasticity of the new tissue. Together, these collagens help create a robust and flexible matrix for new skin and tissue, ensuring that wounds heal properly and maintain their integrity. Increased expression of vascular endothelial growth factor (VEGF) aids in neovascularization and wound healing. By reducing proinflammatory responses and enhancing the regenerative processes of skin cells, lavender oil promotes faster healing, especially in post-surgical and irritated skin conditions.

6. Mental Health and Sleep Improvement

Molecular or Cellular Target/Pathway: Oxytocin and cortisol levels, neuropeptide regulation, melatonin levels.

Mechanisms: The calming effects of lavender lead to improved sleep quality and reduced insomnia symptoms in various populations, including postmenopausal women and individuals experiencing anxiety. Lavender inhalation increases oxytocin production, which may enhance social behaviors and mood, while it lowers cortisol levels linked to stress. Lavender's effects on neurotransmitter systems support its role in mental health. Neuropeptide regulation is closely linked to mental health because these small protein-like molecules act as neurotransmitters and neuromodulators in the brain, influencing mood, behavior, and overall mental well-being. Different neuropeptides, such as substance P, oxytocin, and neuropeptide Y, have been shown to play roles in stress response, emotional regulation, and social behavior. Additionally, lavender oil significantly elevates blood melatonin levels when inhaled.

7. Pain Relief

Molecular or Cellular Target/Pathway: GABA receptors, proinflammatory mediators, ion channels, neurotransmitter receptors.

Mechanisms: Lavender oil has shown efficacy in reducing pain perception in various types of acute and chronic pain models. Specifically, research indicates that lavender oil may modulate the activity of

GABA (gamma-aminobutyric acid) receptors. GABA is an inhibitory neurotransmitter that plays a crucial role in calming the nervous system and reducing pain signals. It also inhibits inflammatory mediators and influences the perception of pain by affecting the brain's processing centers, triggering calm that prevents exacerbation of pain perception. Lastly, linalool is believed to interact with ion channels and neurotransmitter receptors in the nervous system, contributing to its pain-relieving effects. Therefore, the oil modulates neurotransmission in pain pathways, inhibiting excitatory signals while promoting inhibitory responses that reduce overall pain intensity.

8. Endocrine and Hormone Effects

Molecular or Cellular Target/Pathway: Hormone levels.

Mechanisms: Lavender oil likely impacts hormone levels and promotes balance, making it potentially useful in managing symptoms of menopause and diminishing blood pressure levels. Menopausal women experienced significant improvements in hot flashes, vertigo, fatigue, muscle pain, headache, heart palpitations, sexual function, vaginal dryness, and abnormal skin sensations.

9. Intestinal Health

Molecular or Cellular Target/Pathway: Gut microbiome.

Mechanisms: Lavender oil supports beneficial bacteria while inhibiting pathogenic strains, thus aiding in gut health maintenance. The selective destruction of harmful bacteria facilitates the maintenance of a healthy microbiome and prevents dysbiosis, which is linked to various chronic health issues.

10. Skin Health, Restoration, and Protection

Molecular or Cellular Target/Pathway: Collagen I and II, skin cells, skin barrier function.

Mechanisms: Lavender oil encourages collagen synthesis and supports skin barrier function. This indicates that lavender oil can support skin repair and regeneration. Its antioxidant properties reduce oxidative stress on skin cells, ultimately promoting overall skin health and reducing signs of aging.

11. Immune System Modulation

Molecular or Cellular Target/Pathway: Indoleamine 2,3-dioxygenase (IDO) enzyme function, mast cell degranulation.

Mechanisms: Lavender oil constituents may modulate IDO activity, influencing immune function by affecting tryptophan metabolism. IDO is an enzyme that plays a significant role in the regulation of immune function, particularly within the context of immune tolerance and response. IDO helps break down a substance called tryptophan, which is an amino acid that our bodies need. Tryptophan is important for our immune cells, including T cells, which are key players in defending our body against infections. By breaking down tryptophan, IDO can limit how much of it is available for immune cells. Without enough tryptophan, these immune cells can't grow and work as well, which helps keep our immune response in check. This is useful in situations where we need to avoid over-reaction of the immune system, such as during pregnancy. IDO is like a gatekeeper for our immune system. It helps control how strong or weak our immune responses are by managing the amount of important nutrients available and encouraging the right type of immune cells to maintain balance within the body. Normalizing mast cell degranulation can reduce the release of inflammatory mediators, potentially alleviating allergic reactions. This can lead to a balanced immune response, reduced inflammation, and potential allergy relief. Broadly, by targeting these pathways, lavender oil may help regulate the immune system, preventing excessive or inappropriate immune responses.

12. Respiratory Health

Molecular or Cellular Target/Pathway: MUC5b regulation, T helper cytokine modulation.

Mechanisms: Constituents of lavender influence MUC5b expression, affecting mucus viscosity and clearance. By modulating T helper cytokines, lavender oil can impact airway inflammation and constriction. MUC5b is a mucus protein that helps trap and remove bacteria and debris from the airways, but too much can lead to clogged lungs and infections, while too little can reduce the body's ability to clear pathogens. T helper cytokines (Th1, Th2, Th17, and Treg) control the immune response—balancing them prevents excessive inflammation that can damage lung tissue (like in asthma or chronic

infections) while ensuring the body can still fight off viruses and bacteria. Properly modulating these factors supports clear airways, efficient immune defense, and reduced lung inflammation, which is key for better respiratory health.

13. Antioxidant Activity

Molecular or Cellular Target/Pathway: Endogenous antioxidants (SOD, GPx).

Mechanisms: Lavender oil may enhance the activity of antioxidant enzymes like GPx and SOD, which neutralize free radicals. By boosting antioxidant defenses, lavender oil can protect cells and tissues from oxidative damage.

14. Metabolic Health

Molecular or Cellular Target/Pathway: HMG-CoA reductase, sterol regulatory element-binding protein-2 (SREBP-2) activity, ubiquitin-dependent degradation of HMGCR, PPAR-α, mevalonate pathway, liver tissue.

Mechanisms: Linalool helps lower cholesterol and triglycerides by targeting key enzymes and receptors involved in fat metabolism. It reduces cholesterol production by breaking down an enzyme needed for its synthesis and boosts fat burning by activating a protein that tells the body to use stored fats for energy. Linalool slows down the HMG-CoA reductase (HMGCR) enzyme, which is the primary enzyme in your liver that produces cholesterol. This reduces cholesterol production. Ubiquitin-dependent degradation of HMGCR is like the body's garbage disposal system for cholesterol production. It tags unnecessary HMGCR enzymes for removal, preventing the excess production of cholesterol. Linalool boosts this process, helping to eliminate and lower cholesterol. SREBP-2 acts as a switch that turns on cholesterol production in the liver. When the body needs more cholesterol, SREBP-2 tells the genes to make more HMGCR. Linalool helps turn this switch down, keeping cholesterol levels in check. PPAR-α is like the fat-burning thermostat of the body. When PPAR-α is activated, the body burns more fat for energy instead of storing it. Linalool turns up this thermostat, reducing triglycerides (fat in the blood) and helping to prevent obesity and metabolic issues. These effects make linalool a promising natural compound for heart health and metabolic balance, potentially working alongside existing cholesterol-lowering medications. Protecting the liver against oxidative damage supports its detoxification function.

15. Renal and Urinary Health

Molecular or Cellular Target/Pathway: Kidney tissue.

Mechanisms: Lavender oil may protect kidney cells from oxidative stress, supporting renal function. This protects the kidneys from damage and preserves their function.

16. Reproductive Health

Molecular or Cellular Target/Pathway: Spermatogenesis, hormones (testosterone, luteinizing hormone).

Mechanisms: Lavender oil may positively influence spermatogenesis, affecting fertility and hormone production (testosterone, luteinizing hormone). This suggests a potential role in male reproductive health.

17. Cardiovascular Health

Molecular or Cellular Target/Pathway: Myocardium.

Mechanisms: Lavender oil may reduce damage to the heart muscle during a heart attack. When a heart attack happens, part of the heart muscle doesn't get enough blood and gets damaged. Lavender oil helps to reduce this damage. During a heart attack, the heart's ability to pump blood is reduced. Lavender oil helps to improve the heart's pumping action and blood flow. Heart attacks can cause harmful substances called "free radicals" to form, which can further damage the heart. Lavender oil acts like an antioxidant, which means it helps to neutralize these harmful substances and lessen their negative effects. After a heart attack, the heart tissue becomes inflamed, which can worsen the damage. Lavender oil helps to reduce this inflammation. Lavender oil protects the heart by reducing damage from free radicals and inflammation, and by helping the heart to function better during and after a heart attack.

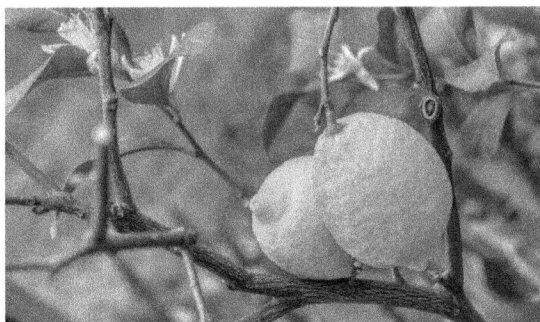

LEMON

Lemon essential oil (*Citrus limon*), often heralded for its diverse therapeutic potential, serves as a vibrant ally in both wellness and beauty routines. It can be obtained by cold pressing or steam distillation of the peels, with the latter having fewer phototoxic furanocoumarins. Among its myriad benefits, lemon oil exhibits promising anticancer activity by activating caspases that induce apoptosis in cancer cells, including those associated with astrocytoma, while sparing normal cells. The oil's antioxidant properties play a crucial role in skin health, helping to inhibit elastase to reduce wrinkle formation and promote a youthful appearance. Furthermore, lemon essential oil supports cardiovascular health by modulating the autonomic nervous system to lower blood pressure and relieve stress. Its neuroprotective effects enhance mood and cognitive function by influencing dopamine and serotonin levels, while also showcasing impressive antimicrobial properties against various pathogens. From promoting digestive health to reducing inflammation and oxidative stress, lemon essential oil proves to be a versatile staple in natural healing, offering a refreshing approach to holistic health that can uplift both body and mind.

SUGGESTED THERAPEUTIC USE:

Cancer: Astrocytoma, cancer cell proliferation

Skin Protection: Wrinkles, loss of skin elasticity, oxidative skin damage

Cardiovascular Conditions: Hypertension, high blood pressure

Neurological Issues: Anxiety, depression, cognitive decline, memory impairment

Infections: Bacterial infections, fungal infections, drug-resistant infections

Gastrointestinal Conditions: Dental caries, heartburn, GERD, digestive issues, gallstones

Inflammatory Conditions: General inflammation, chronic inflammatory conditions (rheumatoid arthritis, IBD, psoriasis, COPD, asthma, lupus, eczema, gout, chronic sinusitis, fibromyalgia, chronic fatigue syndrome, periodontitis)

Organ and Tissue Protection: Oxidative stress-related conditions (cancer, IBD, diabetes, neurodegenerative conditions, cardiovascular disease, liver or kidney disease, metabolic syndrome, fibromyalgia, chronic fatigue syndrome, skin aging), neurodegenerative diseases

Mental Health Concerns: Anxiety, stress, examination stress, medical procedure anxiety

Metabolic Disorders: High cholesterol, impaired glucose metabolism

Pain: Acute and chronic pain, inflammatory pain

Oral Health: Cavities, dental plaque, bad breath

MOLECULAR AND CELLULAR TARGETS, BIOLOGICAL PATHWAYS, AND MECHANISMS:

1. Anticancer Activity

Molecular or Cellular Target/Pathway: Caspases and cancer cell proapoptotic pathways

Mechanisms: Lemon essential oil activates caspases, which are critical for initiating apoptosis in cancer cells, including astrocytoma cells. Caspases are a family of cysteine proteases that play a critical role in apoptosis (programmed cell death), which is essential for maintaining healthy tissue homeostasis and eliminating damaged or potentially harmful cells. In the context of cancer, the regulation of caspases can influence tumor development and progression. If caspases are not activated properly, cancer cells can become resistant to the signals that typically trigger cell death, contributing to tumor growth and metastasis. Additionally, the deregulation of caspases and other apoptotic pathways can lead to inflammatory responses in the tumor microenvironment, further supporting cancer progression. The oil promotes the intrinsic apoptosis signaling pathway, effectively inducing cell death in tumor cells. Its constituents (such as limonene) may enhance apoptosis while sparing normal cells, making it a potential therapy in oncology.

2. Skin Health and Anti-Aging Effects

Molecular or Cellular Target/Pathway: Elastase, oxidative damage.

Mechanisms: Lemon oil inhibits elastase, reducing the breakdown of elastin, which is vital for maintaining skin elasticity. By reducing elastase activity, lemon oil may help preserve the skin's youthful appearance and mitigate wrinkles. Its antioxidant properties help counter oxidative damage, contributing to improved skin health.

3. Cardiovascular Effects

Molecular or Cellular Target/Pathway: Autonomic nervous system activity.

Mechanisms: Inhalation of lemon oil influences autonomic nervous system (ANS) activity, impacting blood pressure. Specifically, it modulates the sympathetic and parasympathetic responses in individuals with hypertension. By reducing stress and promoting relaxation, lemon oil can lower both systolic and diastolic blood pressure, contributing to cardiovascular health.

4. Neuroprotective and Mood-Enhancing Effects

Molecular or Cellular Target/Pathway: Dopamine and serotonin receptors, GABAergic modulation.

Mechanisms: Lemon oil influences neurotransmitter systems linked to mood regulation and cognitive function. By elevating dopamine and serotonin levels, lemon oil helps reduce symptoms of anxiety and depression, and its inhalation has been shown to improve cognitive performance and memory in animal studies. Its interaction with GABA receptors may help alleviate anxiety and promote relaxation. Specifically, lemon oil interacts with GABA(A) receptors, which are vital for calming brain activity and maintaining proper neurological function.

5. Antimicrobial Properties

Molecular or Cellular Target/Pathway: Pathogen cell structure and wall/membrane integrity, biofilm formation.

Mechanisms: Lemon oil shows efficacy against several pathogens, including drug-resistant strains. The antimicrobial activity of lemon essential oil is enhanced by its constituents, such as limonene, which are effective against various bacterial and fungal species. It may disrupt bacterial membrane integrity and prevent biofilm formation.

6. Gastrointestinal Benefits

Molecular or Cellular Target/Pathway: Gastric mucus production, tight junction proteins (occludin, claudin-1, and ZO-1— at both the transcriptional and translational levels), CytoMix-induced paracellular permeability, β-glucose and 2-succinamate reduction, CBR-1 antagonism, immune and inflammatory responses.

Mechanisms: The limonene in lemon oil can reduce symptoms of heartburn and gastroesophageal reflux disease (GERD) without decreasing stomach acid. It increases gastric mucus production and regulates immune and inflammatory responses. It also improved the function of tight junction proteins like occludin, claudin-1, and ZO-1, which are critical for maintaining the intestinal barrier. This improvement helps to reduce gut permeability (leaky gut) and prevent harmful substances from leaking out of the intestine. Lemon oil, due to its rich limonene content, can improve gastrointestinal health by influencing how certain substances pass through the gut and how our body processes sugars. It interacts with something called CytoMix, which can increase the permeability of the gut lining, allowing unwanted particles to leak into the bloodstream. By reducing this permeability, limonene helps maintain a healthier barrier in the intestines. Additionally, it can lower levels of specific sugars like β-glucose and 2-succinamate, which can be harmful in excess. Lastly, limonene acts on CBR-1, a receptor in our gut that, when blocked, can help reduce discomfort and inflammation. Overall, these actions help keep our digestive system running smoothly and prevent issues like bloating or irritation.

7. Inflammation Reduction

Molecular or Cellular Target/Pathway: 5-LOX, cytokine production regulation, cellular migration, and protein extravasation.

Mechanisms: Lemon essential oil inhibits 5-LOX activity, which is pivotal in the inflammatory response. The oil reduces the production of proinflammatory cytokines, contributing to its anti-inflammatory effects. Lemon oil has demonstrated the capacity to reduce cellular migration and protein extravasation, which are key processes in the inflammatory response. Cellular migration and protein extravasation are key processes in the body's response to inflammation. When there is tissue damage or infection, immune cells leave the bloodstream and move to the affected area—a process called cellular migration. During this process, proteins in the blood, such as antibodies and clotting factors, exit the vessels (extravasation) to help fight infection or heal damage. This movement of cells and proteins causes redness, swelling, and pain, which are hallmark signs of inflammation. Essentially, these processes help deliver the immune system to the site of injury or infection, making them crucial for the body's healing response.

8. Oxidative Stress Protection

Molecular or Cellular Target/Pathway: Oxidative stress-inducing agents, endogenous antioxidant enzyme activity.

Mechanisms: Lemon essential oil protects against oxidative stress in various organ systems. It boosts activities of antioxidant enzymes, such as SOD, which help mitigate oxidative damage. By reducing lipid peroxidation and promoting neuroprotection, lemon essential oil may aid in the prevention of neurodegenerative diseases. Limonene also maintains and restores healthy liver glutathione levels.

9. Anxiolytic and Adaptogenic Effects

Molecular or Cellular Target/Pathway: Autonomic nervous system activity, norepinephrine and GABA activity, corticosterone levels.

Mechanisms: Inhalation of lemon oil induces significant changes in corticosterone levels and balances ANS activity. Its calming properties help diminish anxiety responses, particularly in stressful situations such as examinations or medical procedures. The oil modulates levels of norepinephrine and enhances GABAergic transmission to reduce anxiety. Norepinephrine is a neurotransmitter and hormone that plays a key role in the body's "fight or flight" response to stress. In situations of stress or anxiety, norepinephrine levels can increase, which leads to heightened arousal, alertness, and vigilance. This increase can enhance feelings of anxiety and can contribute to the physiological symptoms of anxiety disorders, such as increased heart rate, sweating, and hyperarousal. Additionally, imbalances in norepinephrine levels may play a role in the development or exacerbation of anxiety disorders.

10. Metabolism, Cardiovascular Function, and Pain

Molecular or Cellular Target/Pathway: PPAR receptors, liver X receptor beta signaling

Mechanisms: Lemon oil influences blood lipid profiles and metabolic parameters positively. Activation of specific nuclear receptors (PPARs and LXRs) for metabolic regulation and the binding to opioid receptors for pain relief, demonstrating a multifaceted approach to health benefits. Limonene activates Peroxisome proliferator-activated receptors and liver X receptor beta signaling to balance blood sugar and improve cholesterol levels. PPARs and liver X receptors (LXRs) are nuclear receptor proteins that play critical roles in the regulation of metabolism, including glucose and lipid metabolism. By activating PPARs, particularly PPAR-gamma and PPAR-alpha, there is an enhancement of glucose uptake and metabolism in peripheral tissues, along with improved lipid profiles due to increased fatty acid oxidation and decreased circulating triglycerides. LXR activation increases the expression of genes involved in cholesterol efflux, such as ATP-binding cassette transporters. This helps to remove excess cholesterol from cells and transport it to the liver for excretion (bile), thereby reducing overall cholesterol levels. LXRs also regulate lipid synthesis and metabolism. By promoting pathways that enhance fatty acid synthesis and storage, LXRs can support stable energy storage and utilization, which helps to manage blood sugar levels. They can also influence insulin sensitivity positively.

11. Oral Health

Molecular or Cellular Target/Pathway: Glucosyltransferase, lactate dehydrogenase activity, acid production in the mouth.

Mechanism: Lemon oil inhibits enzymes—glucosyltransferase and lactate dehydrogenase—that are linked to the development of dental caries. Glucosyltransferase helps bacteria in the mouth produce sticky substances that cause plaque and cavities. Lemon oil has natural antibacterial properties, which can inhibit the activity of glucosyltransferase, making it harder for harmful bacteria to thrive. It also affects lactate dehydrogenase, an enzyme involved in the production of acid by these bacteria. Since excess acid can erode tooth enamel and lead to cavities, lemon oil's interaction helps reduce acid

production as well. By lowering both plaque buildup and acid levels, lemon oil contributes to healthier teeth and gums. However, it's important to use lemon oil properly, as too much acidity in the oil itself can also harm teeth if not diluted.

12. Pain

Molecular or Cellular Target/Pathway: Opioid receptors, antioxidant activity.

Mechanism: Lemon oil exhibits antinociceptive activity, meaning it reduces the perception of pain. The antinociceptive effect of lemon oil appears to involve central inhibitory mechanisms, possibly related to the opioid system. By binding to opioid receptors, lemon oil can reduce pain. Lemon oil has demonstrated a strong antioxidant potential by effectively scavenging free radicals. Free radicals are involved in pathological processes including pain. By reducing free radicals, lemon oil can alleviate pain. Specifically, lemon oil prevents oxidative damage to lipids, which is a key process in inflammation and pain.

13. Gallstones

Molecular or Cellular Target/Pathway: Enhanced bile acid metabolism and bile composition, solubilization of cholesterol.

Mechanism: Limonene, a compound found in citrus oils, has been studied for its potential impact on gallstones, particularly in terms of dissolution. Gallstones are often composed of cholesterol. Limonene may increase the solubility of cholesterol in bile, potentially preventing the crystallization of cholesterol that leads to gallstone formation or promoting the dissolution of existing stones. Limonene has been shown to stimulate bile flow, which can help in dissolving gallstones. Increased bile flow can facilitate the breakdown and removal of cholesterol, thus aiding in the dissolution of stones. Some animal studies have suggested that limonene can help reduce the size of gallstones or prevent their formation by altering bile composition and enhancing the body's natural bile acid metabolism.

LEMONGRASS

Lemongrass essential oil (*Cymbopogon flexuosus, C. citratus*) is a powerful oil that provides a wide array of health benefits, making it an essential component for natural wellness and skincare. This invigorating oil showcases impressive anticancer activity by inducing apoptosis in various cancer cell lines, including lung sarcoma and neuroblastoma, through the activation of intrinsic and extrinsic apoptotic pathways—crediting its constituents, particularly citral, with the inhibition of cellular proliferation and tumor growth. In addition to its anticancer potential, lemongrass oil boasts broad-spectrum antimicrobial properties, effectively combating pathogens like *C. albicans* and *S. aureus* by disrupting cell membranes and inhibiting biofilm formation. Its anti-inflammatory effects help balance cytokine production, offering relief from chronic inflammatory conditions, while its neuroprotective activities support brain health by modulating GABAergic activity and enhancing antioxidant defenses. Furthermore, lemongrass essential oil excels in gastrointestinal protection, metabolic regulation, and even serves as a natural insect repellent, showcasing its versatility and efficacy across various domains of health. Whether used for enhancing personal care or as a therapeutic agent, lemongrass oil stands out as a holistic remedy with remarkable therapeutic potential.

SUGGESTED THERAPEUTIC USE:

Cancer: Lung sarcoma, colon cancer, neuroblastoma, promyelocytic leukemia

Infections: Fungal infections, bacterial infections, multidrug-resistant infections, dental infections

Inflammatory Conditions: Chronic inflammatory conditions (rheumatoid arthritis, IBD, psoriasis, COPD, asthma, lupus, eczema, gout, chronic sinusitis, fibromyalgia, chronic fatigue syndrome, periodontitis)

Neuroprotection: Glutamate toxicity, seizures, neurodegenerative diseases, neuroinflammation

Pain: General pain, inflammatory pain, neurogenic pain

Gastroprotection: Gastric ulcers, gastrointestinal disorders

Organ and Tissue Protection: Oxidative stress-related conditions (cancer, IBD, diabetes, neurodegenerative conditions, cardiovascular disease, liver or kidney disease, metabolic syndrome, fibromyalgia, chronic fatigue syndrome, skin aging), aging, cellular damage

Metabolic Disorders: Diabetes, hyperglycemia, impaired lipid metabolism

Insect Control and Repellency: Mosquito infestations, agricultural pests

Skin Issues: Skin infections, fungal skin conditions (e.g., candida, dermatophyte infections)

MOLECULAR AND CELLULAR TARGETS, BIOLOGICAL PATHWAYS, AND MECHANISMS:

1. **Anticancer Activity**

Molecular or Cellular Target/Pathway: Caspases, cancer cell intrinsic and extrinsic proapoptotic pathways.

Mechanisms: Lemongrass essential oil induces apoptosis in various cancer cell lines, including lung sarcoma, colon, neuroblastoma, and promyelocytic leukemia. Constituents like citral are credited with triggering apoptosis, inhibiting cellular proliferation, and reducing tumor growth in preclinical models. The oil activates both internal signaling (mitochondrial pathway) and external signaling pathways, leading to programmed cell death. When a cell is under stress or damaged, it can trigger a process called programmed cell death, or apoptosis, to protect the body from harmful cells, like cancer. The internal signaling pathway starts when certain signals inside the damaged cell cause tiny structures in the cell called mitochondria to release a molecule called cytochrome c. This molecule then activates special proteins known as caspases, which are like the cell's executioners—they help break down the cell's components to safely dispose of them. There's also an external signaling pathway that works differently. This one involves proteins on the cell's surface called death receptors. When these receptors bind to specific molecules (ligands), they send signals inside the cell that activate the same caspases, leading to cell death. In short, both the internal and external pathways lead to the same process: they activate caspases that systematically destroy the cell's parts, effectively leading to the cell's death and helping the body get rid of potentially dangerous cells, like cancer cells.

2. **Antimicrobial Properties**

Molecular or Cellular Target/Pathway: Pathogen cell structures and cell wall/membrane integrity, biofilm formation.

Mechanisms: Lemongrass essential oil demonstrates broad-spectrum antimicrobial activity against various pathogens. Its constituents prevent the growth of multidrug-resistant strains *in vitro*, suggesting its potential as a topical agent for infections and in dental applications. The oil inhibits biofilm formation and disrupts the integrity of pathogenic cell membranes, leading to cell death.

3. Anti-inflammatory Effects

Molecular or Cellular Target/Pathway: Proinflammatory cytokines (IL-1β and IL-6), NLRP3 inflammasome modulation.

Mechanisms: By modulating key inflammatory mediators, lemongrass essential oil can relieve symptoms of chronic inflammatory conditions. Lemongrass oil inhibits the production of cytokines such as IL-1β and IL-6, which are involved in inflammatory responses. The oil promotes a balanced activation of the NLRP3 inflammasome, which is key in resolving inflammation in various diseases. The NLRP3 inflammasome is a group of proteins that act like an alarm system inside your cells that triggers inflammation when it senses danger. Inflammation is a normal and necessary part of your body's defense against infections and injuries. However, if the NLRP3 inflammasome gets overactive, it can lead to too much inflammation, which can contribute to various diseases (metabolic disorders, neurodegeneration, autoinflammatory conditions, cardiovascular disease, IBD, osteoporosis, cancer, and the severity of COVID-19).

4. Neuroprotective Effects

Molecular or Cellular Target/Pathway: Glutamate and GABA receptors, antioxidant pathways.

Mechanisms: The oil mitigates neuroinflammation and damage following exposure to neurotoxic agents, making it possibly beneficial for conditions like seizures and neurodegenerative diseases. Lemongrass oil modulates GABAergic activity, reducing neuronal excitability and providing neuroprotection against glutamate toxicity. Glutamate toxicity refers to the detrimental effects caused by excessive levels of the neurotransmitter glutamate in the brain, which can lead to neuronal damage and cell death. Although most often associated with conditions like stroke, traumatic brain injury, and neurodegenerative diseases, glutamate toxicity can also be triggered by brain inflammation, drugs (e.g., antidepressants), MSG, excess caffeine, heavy metals, and alcohol. The oil also increases antioxidant defenses and decreases oxidative stress, which is pivotal in protecting brain health.

5. Painful Conditions

Molecular or Cellular Target/Pathway: Opioid receptors, central and peripheral nervous system activity and receptors, prostaglandin synthesis, and histamine/serotonin-mediated inflammation.

Mechanisms: It exhibits pain-relieving effects through multiple mechanisms, including activating opioid receptors (as evidenced by naloxone blockage), inhibiting prostaglandin synthesis and histamine/serotonin-mediated inflammation, and modulating pain signals in both neurogenic and inflammatory pathways. Lemongrass oil influences pathways that regulate pain perception. Pain relief provided by lemongrass is blocked by naloxone, suggesting involvement of opioid receptors. The oil also inhibited prostaglandin synthesis, a key pathway in pain and inflammation, and histamine/serotonin-mediated inflammation, which are involved in pain signaling.

Additionally, lemongrass reduced pain in both neurogenic and inflammatory pain models, meaning it acts at both the nervous system (brain/spinal cord) and inflammatory sites. Researchers attribute the pain-relieving effects of lemongrass oil to its citral and 6-methylhepta-5-en-2-one content. These constituents are thought to inhibit prostaglandin biosynthesis, similar to how non-steroidal anti-inflammatory drugs work.

6. Gastrointestinal Protection

Molecular or Cellular Target/Pathway: Prostaglandin production.

Mechanisms: The oil protects against ethanol-induced ulcers and promotes healing, thus demonstrating therapeutic potential for gastrointestinal disorders. Lemongrass modulates prostaglandin synthesis, which is integral to mucosal protection in the stomach. Prostaglandins are natural substances produced by the body that play a key role in protecting the stomach and intestines. They help maintain a thick layer of mucus in the stomach lining, which acts like a shield against stomach acid and harmful substances. Additionally, prostaglandins promote blood flow to the stomach, which helps support the healing of any injuries or damage. They also reduce the production of stomach acid, which further protects the digestive tract.

7. Antioxidant Activity

Molecular or Cellular Target/Pathway: ROS, cellular antioxidant defenses.

Mechanisms: Lemongrass essential oil effectively scavenges free radicals and mitigates oxidative stress. By enhancing cellular antioxidant defenses, the oil may protect against oxidative damage, which is linked to aging and various diseases.

8. Metabolic Regulation

Molecular or Cellular Target/Pathway: Insulin signaling pathways (PTP-1B, PPAR-γ).

Mechanisms: The oil's constituents facilitate better glucose control and lipid metabolism, suggesting its potential in managing diabetes. Lemongrass promotes improved insulin sensitivity by inhibiting PTP-1B and activating PPAR-γ. PTP-1B and PPAR-γ are two important players in how our body manages blood sugar (glucose) and responds to insulin, which is the hormone that helps cells take in glucose. When PTP-1B is inhibited (blocked), it helps to boost the action of insulin, making your body more sensitive to it. This means that insulin can work better to lower blood sugar levels. On the other hand, activating PPAR-γ helps improve how the body handles fats and sugars, promoting insulin sensitivity as well. Together, inhibiting PTP-1B and activating PPAR-γ can lead to better glucose control, making it easier for your body to keep blood sugar levels balanced and functioning effectively. This can be especially beneficial for people at risk for diabetes, or with metabolic syndrome or insulin resistance.

9. Insecticidal and Repellent Properties

Molecular or Cellular Target/Pathway: Insect nervous systems.

Mechanisms: Lemongrass exhibits toxicity to various pests, affecting nervous function and survival, by interfering with nervous system function. The oil acts as a natural insecticide and repellent, effective against mosquitoes and agricultural pests, making it valuable for pest control.

10. Dermatological Applications

Molecular or Cellular Target/Pathway: Skin pathogens.

Mechanisms: Lemongrass oil shows efficacy against skin infections and fungal conditions. Its antifungal and antimicrobial properties allow it to treat various skin disorders, including candida and dermatophyte infections.

LIME

Lime (*Citrus aurantifolia*) essential oil presents a diverse range of therapeutic properties, evidenced by both *in vitro* and *in vivo* studies. It can be distilled or cold pressed, with the distilled form being more complex in composition and sweeter in aroma. Its applications extend from metabolic regulation to antimicrobial defense, highlighting its multifaceted potential. Research indicates that lime oil can promote weight loss and prevent obesity, inhibit pathogenic fungi and bacteria, and

demonstrate anticancer activity by inducing apoptosis. Furthermore, it exhibits significant antioxidant capabilities and modulates enzyme activity, notably inhibiting acetylcholinesterase and butyrylcholinesterase, which are crucial in neurodegenerative contexts. Lime oil also demonstrates anti-inflammatory effects by reducing cytokine production and white blood cell migration, and exhibits activity against parasites. These findings underscore the complex interplay of lime oil's constituents with various biological systems.

SUGGESTED THERAPEUTIC USE:

Metabolic Disorders: Obesity, type 2 diabetes, high cholesterol

Neurological Conditions: Alzheimer's disease, neurodegenerative conditions

Infections: Bacterial, fungal, parasitic

Inflammatory Conditions: Muscle soreness, general inflammation

Cardiovascular Disease: Atherosclerosis, high blood pressure

Skin Conditions: Skin infections, wounds

Oral Health: Dental cavities

Mental Health Issues: Anxiety, stress, mood disturbances

Pest Control: Insect repellency and insecticide

MOLECULAR AND CELLULAR TARGETS, BIOLOGICAL PATHWAYS, AND MECHANISMS:

1. Weight Loss and Obesity Prevention

Molecular or Cellular Target/Pathway: Lipid metabolism, glucose homeostasis.

Mechanisms: Lime essential oil modulates hepatic (liver) glucose levels, improves lipid profiles (reduction of LDL, increase of HDL), and reduces triglycerides. Specifically, the oil increased excretion of neutral sterols in the feces. Not all of the cholesterol produced by the body is used. Some of it must be eliminated. One way your body gets rid of cholesterol is by converting it into other substances, like neutral sterols. These neutral sterols can't be easily absorbed back into your body. So, when you excrete more of them in your feces (poop), you're essentially helping your body eliminate cholesterol. This can lower the amount of cholesterol circulating in your blood, which is a good thing for maintaining healthy cholesterol levels.

2. Antifungal Activity

Molecular or Cellular Target/Pathway: Fungal cell wall/membrane integrity, ergosterol biosynthesis.

Mechanisms: Lime oil disrupts fungal cell wall and membrane integrity and interferes with ergosterol biosynthesis, thereby inhibiting fungal growth.

3. Anticancer Activity

Molecular or Cellular Target/Pathway: Cancer cell proapoptotic pathways, cell cycle regulation.

Mechanisms: Lime oil induces apoptosis in cancer cells and inhibits cancer cell proliferation.

4. Neuroprotective Activity

Molecular or Cellular Target/Pathway: Cholinergic system, AChE, BChE.

Mechanisms: Inhibiting both AChE and BChE, lime oil increases acetylcholine levels, which are crucial for neurological function. The oil also possesses antioxidant activity, which may reduce free radical damage in nerve cells.

5. Antibacterial Activity

Molecular or Cellular Target/Pathway: Bacterial cell wall/membrane integrity, cytoplasmic membrane function.

Mechanisms: Lime oil disrupts bacterial cell membranes and interferes with bacterial enzyme function, leading to inhibition of bacterial growth.

6. Anti-inflammatory Activity

Molecular or Cellular Target/Pathway: Cytokine production, leukocyte migration, iNOS, protein extravasation, NF-κB, reactive nitrogen species (RNS).

Mechanisms: Lime oil promotes a reduction of cytokine release (TNF-α, IL-1β, IFN-γ) by interfering with NF-κB signaling. It also inhibits leukocyte (polymorphonuclear neutrophils and mononuclear cells) cell migration and reduces protein extravasation. When inflammation occurs, blood vessels become more permeable, allowing proteins and fluid to leak out into surrounding tissues. This process, called protein extravasation, contributes to swelling, redness, and pain by increasing fluid buildup and activating immune cells to fight infection or injury. The oil also decreases excess nitric oxide production, which suppresses reactive nitrogen species (RNS) production and oxidative stress. RNS are unstable molecules that the body produces during inflammation. While they help fight infections, too much RNS can damage healthy cells, leading to more inflammation, pain, and even tissue injury.

7. Antidiabetic Activity

Molecular or Cellular Target/Pathway: Glucose metabolism, glycogen metabolism.

Mechanisms: Lime leaf essential oil reduces fasting blood glucose, increases liver glycogen levels, and improves lipid profiles. The oil increased glucose utilization and improved storage of glucose as glycogen. Based on research with limonene, it is possible that limonene regenerated pancreatic beta-cells, stimulated insulin secretion, and adjusted carbohydrate-metabolizing enzymes.

8. Antioxidant Activity

Molecular or Cellular Target/Pathway: Free radical scavenging, oxidative stress pathways.

Mechanisms: Lime oil donates electrons to neutralize free radicals, thereby inhibiting oxidative damage.

9. Cardiovascular Disease Prevention

Molecular or Cellular Target/Pathway: Vascular smooth muscle cell (VSMC) proliferation, cell cycle regulation.

Mechanisms: Inhibition of VSMC proliferation, potential for therapeutic intervention in vascular diseases. The abnormal proliferation of VSMCs plays a major role in cardiovascular diseases like atherosclerosis. VSMCs are found in the vascular wall and are responsible for the tone and contractions of arteries, which regulates blood pressure and flow in response to specific metabolic demands. By inhibiting VSMC proliferation, lime essential oil could potentially help prevent or slow down the progression of atherosclerosis and other cardiovascular diseases that involve abnormal blood vessel growth.

MARJORAM (SWEET)

Marjoram essential oil, scientifically known as *Origanum majorana*, is a remarkable natural remedy that boasts a plethora of therapeutic benefits, making it a vital addition to holistic health practices. This versatile oil shows significant anticancer activity by inducing apoptosis across various cancer cell lines, including myeloid leukemia, breast, and prostate cancer, through the activation of intrinsic and extrinsic apoptotic pathways. Its key constituents, such as terpinen-4-ol and sabinene hydrate, enhance survivin regulation, effectively reducing cancer cell viability. Furthermore, marjoram oil excels in its anti-inflammatory effects, inhibiting proinflammatory cytokines and the COX-2 enzyme, which can alleviate symptoms of chronic inflammatory conditions like arthritis and atherosclerosis. The oil's broad-spectrum antimicrobial properties also provide protection against pathogens by disrupting their membrane integrity, while its neuroprotective effects can enhance cognitive function and mitigate oxidative stress. Combined with its supportive roles in cardiovascular health, gastroprotection, and even antiparasitic activity, marjoram essential oil is a multifaceted ally in promoting both physical and mental well-being.

SUGGESTED THERAPEUTIC USE:

Cancer: Myeloid leukemia, breast cancer, prostate cancer

Inflammatory Conditions: Arthritis, atherosclerosis, general inflammation

Infections: Bacterial infections, fungal infections, multidrug-resistant infections, biofilm-related infections (chronic wound infections, catheter-associated UTIs, heart valve infections, chronic sinusitis, cystic fibrosis, dental plaque, periodontal disease, osteomyelitis, implant-associated infections, ventilator-associated pneumonia, conjunctivitis, dialysis-associated infections)

Neuroprotection and Mental Health Issues: Alzheimer's disease, cognitive decline, anxiety

Organ and Tissue Protection: Oxidative stress-related conditions (cancer, IBD, diabetes, neurodegenerative conditions, cardiovascular disease, liver or kidney disease, metabolic syndrome, fibromyalgia, chronic fatigue syndrome, skin aging), aging, cellular damage

Pain: Menstrual pain, arthritis pain, general pain

Cardiovascular Issues: High blood pressure, stress-related cardiovascular issues

Gastroprotection: Gastric ulcers, drug-induced ulcers, gastrointestinal irritation

Parasitic Infections: Malaria

Insect Control: Mosquito infestations, tick infestations

MOLECULAR AND CELLULAR TARGETS, BIOLOGICAL PATHWAYS, AND MECHANISMS:

1. Anticancer Activity

Molecular or Cellular Target/Pathway: Caspases and intrinsic and extrinsic cancer cell proapoptotic pathways.

Mechanisms: Marjoram essential oil induces apoptosis in various cancer cell lines, including myeloid leukemia, breast, and prostate cancer cells. The oil activates internal and external apoptotic signals to promote cell death in cancer cells. The key constituents (terpinen-4-ol, sabinene hydrate, α-terpinene, and γ-terpinene) enhance survivin regulation, which is crucial for modulating apoptosis, and reduce cancer cell viability. Survivin is a protein that plays a key role in helping cells survive and grow, especially during the critical process of cell division. In healthy cells, survivin is tightly regulated, meaning that it is produced when needed and turned off when it's not. However, in many types of cancer, survivin is found in abnormally high levels. This unchecked production allows cancer cells to evade normal processes that would lead to their death, helping tumors grow and spread.

2. Anti-inflammatory Effects

Molecular or Cellular Target/Pathway: Proinflammatory cytokines, COX-2, NF-κB.

Mechanisms: By suppressing inflammatory processes, marjoram oil aids in reducing symptoms of conditions like arthritis and atherosclerosis. Marjoram oil inhibits the production of proinflammatory cytokines (TNF-α, IL-1β, IL-6, IL-10) and the COX-2 enzyme, which is involved in the inflammatory response. It also inhibits the expression of NF-κB and other inflammatory mediators.

3. Antimicrobial Properties

Molecular or Cellular Target/Pathway: Pathogen cell structures and wall/membrane integrity, biofilm formation.

Mechanisms: Marjoram oil exhibits broad-spectrum antimicrobial activity against various bacterial and fungal species. The oil disrupts bacterial and fungal membrane integrity, leading to cell death. The constituents, especially terpenoids, inhibit biofilm formation and exhibit significant action against multidrug-resistant organisms.

4. Neuroprotective and Cognitive Enhancement

Molecular or Cellular Target/Pathway: AChE, GABA receptor activity, oxidative stress.

Mechanisms: Marjoram oil inhibits AChE, potentially increasing acetylcholine levels, which is vital for memory and cognitive function. The oil is also involved in modulating GABA receptor activity, impacting neuronal excitability and anxiety levels. Through its neuroprotective effects, marjoram oil can mitigate oxidative stress and promote cognitive function, particularly shown in studies involving Alzheimer's disease models.

5. Oxidative Stress Reduction

Molecular or Cellular Target/Pathway: ROS, antioxidant pathways.

Mechanisms: Marjoram essential oil demonstrates antioxidant properties that combat oxidative stress. Through the modulation of antioxidant pathways and direct scavenging of free radicals, marjoram can mitigate cellular damage caused by oxidative agents.

6. Reduction of Pain and Inflammation

Molecular or Cellular Target/Pathway: Pain-associated neurotransmitters, COX-2, leukotriene synthesis, direct relaxant effect on muscles, calcium influx.

Mechanisms: Its relaxing effects on muscles and modulation of pain-associated neurotransmitters help alleviate conditions such as menstrual pain and arthritis. Marjoram oil influences various pathways involved in pain perception and inflammation, including COX-2 and leukotriene synthesis. Its primary constituent, terpinen-4-ol, possesses antispasmodic activity in the GI tract, suggesting it may be helpful for digestive pain. Terpinen-4-ol appears to inhibit calcium influx through voltage-gated calcium channels or effects on intracellular calcium stores. It also has a direct effect on smooth muscle cells.

7. Cardiovascular Benefits

Molecular or Cellular Target/Pathway: Cortisol, autonomic nervous system function.

Mechanisms: By influencing the autonomic nervous system (ANS) and reducing stress, marjoram essential oil can promote cardiovascular health and contribute to relaxation. Inhalation of marjoram oil in a blend with ylang ylang, lavender, and neroli is associated with decreases in cortisol levels and systolic blood pressure.

8. Gastroprotective Effects

Molecular or Cellular Target/Pathway: Mucosal defense mechanisms.

Mechanisms: It helps mend the stomach lining and enhances mucosal defenses against irritants like ethanol and NSAIDs. Marjoram oil has been found to replenish lost gastric mucous and protect against drug-induced ulcers. Replenishing lost gastric mucus is essential for protecting the stomach because this mucus forms a critical barrier between the stomach lining and harsh stomach acid. If the mucus layer becomes thin or damaged, the stomach is more vulnerable to irritation, ulcers, and even bleeding. By increasing the production of gastric mucus, we can help restore that protective barrier. This ensures the stomach lining stays safe from acid

damage, supporting overall digestive health and preventing gastrointestinal problems.

9. Antiparasitic Activity

Molecular or Cellular Target/Pathway: Plasmodium parasite metabolism or induction of oxidative stress.

Mechanisms: Marjoram essential oil appears to exert its antimalarial effects through a combination of mechanisms, likely involving the disruption of parasite metabolism and the induction of oxidative stress. Marjoram essential oil exhibits activity against malarial parasites while preserving healthy cells. Malaria is a disease caused by tiny parasites that are spread to humans through the bites of infected mosquitoes. Once inside the body, these parasites travel to the liver, multiply, and then infect red blood cells. Inside the red blood cells, they continue to grow and multiply, eventually causing the cells to burst. This leads to symptoms like fever, chills, and anemia because the red blood cells are crucial for carrying oxygen.

10. Insecticidal Properties

Molecular or Cellular Target/Pathway: Insect nervous systems.

Mechanisms: Marjoram oil is effective against pests such as mosquitoes and ticks. The oil acts as an insect repellent and insecticide, damaging the nervous systems of insect pests, leading to mortality.

MELISSA

Melissa essential oil, derived from the plant *Melissa officinalis*, is a potent natural remedy renowned for its multifaceted therapeutic properties. Notably, it exhibits significant anticancer activity by inducing apoptosis in various cancer cell lines, including lung carcinoma and glioblastoma multiforme, through the activation of intrinsic apoptotic pathways, which ultimately reduces tumor growth. In addition to its anticancer benefits, melissa oil plays a vital role in neuroprotection by inhibiting enzymes such as acetylcholinesterase, enhancing acetylcholine levels, and thereby supporting cognitive function—a particularly promising application for neurodegenerative conditions like Alzheimer's disease. The oil also demonstrates anxiolytic and sedative effects, aiding in the management of anxiety and agitation, as well as providing potent anti-inflammatory and antimicrobial properties that combat infections and inflammatory responses. With its ability to scavenge free radicals, manage metabolic effects, and even promote cardiovascular health, melissa essential oil is promising to enhance both mental and physical well-being.

SUGGESTED THERAPEUTIC USE:

Cancer: Lung carcinoma, breast cancer, colorectal adenocarcinoma, glioblastoma multiforme

Neuroprotection: Alzheimer's disease, cognitive decline, memory impairment, neurological oxidative stress

Mental Health and Behavioral Issues: Anxiety, agitation, stress

Inflammatory Conditions: Chronic inflammatory conditions (rheumatoid arthritis, IBD, psoriasis, COPD, asthma, lupus, eczema, gout, chronic sinusitis, fibromyalgia, chronic fatigue syndrome, periodontitis), general inflammation

Infections: Viral infections (HSV-1, HSV-2), fungal infections (*C. albicans*)

Organ and Tissue Protection: Oxidative stress-related conditions (cancer, IBD, diabetes, neurodegenerative conditions, cardiovascular disease, liver or kidney disease, metabolic syndrome, fibromyalgia, chronic fatigue syndrome, skin aging), aging, cellular damage

Metabolic Disorders: Type 2 diabetes, hyperglycemia, hypertriglyceridemia

Pain: Chronic pain, neuropathic pain

Parasitic Infection: African sleeping sickness, Acanthamoeba infections

Cardiovascular Issues: Acute coronary syndrome, stress-related cardiovascular issues

MOLECULAR AND CELLULAR TARGETS, BIOLOGICAL PATHWAYS, AND MECHANISMS:

1. **Anticancer Activity**

Molecular or Cellular Target/Pathway: Cancer cell mitochondrial proapoptotic pathways.

Mechanisms: Melissa oil causes apoptosis in various cancer cell lines, including lung carcinoma, breast, colorectal adenocarcinoma, and glioblastoma multiforme. The oil promotes cellular apoptosis through the regulation of gene expression related to cancer progression and survival, effectively reducing tumor growth. It also activates caspases and other apoptotic signals, leading to programmed cell death. Some cancer cells develop resistance to apoptosis, often due to mutations, overexpression of antiapoptotic proteins (like Bcl-2), or downregulation of proapoptotic proteins (like Bax). These alterations can lead to mitochondrial dysfunction, where the mitochondria fail to release signals that would typically trigger apoptosis, allowing cancer cells to survive and proliferate.

2. Neuroprotective Effects

Molecular or Cellular Target/Pathway: AChE, BChE.

Mechanisms: Melissa oil inhibits these enzymes (AChE, BChE), which are involved in breaking down acetylcholine, choline esters, and aliphatic and aromatic esters, crucial for cognitive functions. Enhanced levels of acetylcholine promote memory and cognitive function, making it beneficial in neurodegenerative diseases such as Alzheimer's. The oil reduces neurological oxidative stress, prevents lipid peroxidation, and protects neuronal cells, collectively enhancing cognitive performance.

3. Anxiolytic and Sedative Effects

Molecular or Cellular Target/Pathway: Nicotinic and muscarinic receptors, GABAergic system, cholinergic system.

Mechanisms: Works by gently adjusting the activity of certain "on/off" switches in the brain that control alertness and relaxation, and by changing the release of chemicals that affect those states. Melissa oil interacts with brain receptors that modulate neurotransmitter release and neuronal excitability. Both nicotinic and muscarinic receptors are like "on/off" switches for the chemical acetylcholine in your brain and body. When stimulated, nicotinic receptors are generally stimulatory and increase anxiety, while blocking them may have some calming effects. The role of muscarinic receptors is more varied, with some contributing to calming effects, while others can be involved in things like muscle control. Therefore, the effects of muscarinic receptors on anxiety and sedation are more complex and depend on which subtype of receptor is involved. The oil also reduces anxiety through GABA receptor pathways, reducing arousal and agitation.

4. Anti-inflammatory Properties

Molecular or Cellular Target/Pathway: 5-LOX, prostaglandin synthesis.

Mechanisms: Melissa oil inhibits the 5-LOX enzyme involved in inflammatory processes. Reducing the formation of proinflammatory mediators, like prostaglandins, leads to decreased inflammation and pain. The oil exhibits potent anti-inflammatory action by regulating inflammatory pathways, potentially benefiting conditions characterized by chronic inflammation.

5. Antimicrobial Effects

Molecular or Cellular Target/Pathway: Pathogen replication and viability pathways.

Mechanisms: Melissa oil inhibits the replication of viruses such as HSV-1 and HSV-2, and also shows activity against fungi. It inhibits viral attachment and subsequent replication in host cells. In fungi, melissa oil increases membrane permeability, destroying the external membrane of the fungi, and modifies the cell morphology—shape, structure, and form. The oil disrupts the ability of pathogens to adhere to host cells and invade healthy tissues, making it a potential therapeutic agent for viral and fungal infections.

6. Antioxidant Activity

Molecular or Cellular Target/Pathway: ROS, oxidative stress.

Mechanisms: Melissa essential oil acts as a free radical scavenger, preventing oxidative damage. By mitigating oxidative stress, melissa oil protects cells from damage caused by free radicals, which could contribute to various chronic diseases and aging.

7. Metabolic Effects

Molecular or Cellular Target/Pathway: Glucose metabolism pathways, glucokinase (GCK) activity, inhibiting glucose-6-phosphatase, mRNA expression of hepatic GLUT4 and SREBP-1c and adipocyte GLUT4, PPAR-γ, PPAR-α, and SREBP-1c.

Mechanisms: Melissa oil enhances glucose uptake and prevents gluconeogenesis, impacting diabetic conditions. Gluconeogenesis is a metabolic pathway that results in the generation of glucose from non-carbohydrate carbon substrates such as lactate, glycerol, and glucogenic amino acids. Melissa also affects several other pathways involved in glucose metabolism. First, it enhances the activity of an important enzyme called glucokinase (GCK), which helps convert glucose into a form that can be stored or used for energy, thereby lowering blood sugar levels. At the same time, melissa oil inhibits another enzyme, glucose-6-phosphatase (G6Pase), which normally raises blood sugar by helping produce glucose in the liver. This combination boosts how well the liver processes and uses glucose. Additionally, it appears to activate transport proteins, such as GLUT4, which help move glucose into cells, enhancing overall glucose absorption. Lastly, melissa influences several other proteins involved in fat and carbohydrate metabolism, notably PPAR-γ and SREBP-1c, which help regulate how the body stores and uses fats and sugars. By modulating metabolic pathways, the oil aids in managing type 2 diabetes, potentially improving insulin sensitivity and reducing blood triglycerides.

8. Pain Relief and Analgesia

Molecular or Cellular Target/Pathway: Opioid and serotonin pathways, potassium adenosine triphosphate pathway, TRPV1, iNOS, and NF-kB.

Mechanisms: The oil has been shown to reduce chronic pain and inflammation, especially in neuropathic pain models, by disrupting pain signaling and diminishing inflammatory mediators associated with pain. Melissa oil modifies pathways related to pain (K ATP, opioid, and serotonin pathways) perception and mood regulation. Modulating K ATP channels, opioidergic pathways, and serotonergic pathways helps relieve pain by targeting different mechanisms in our body. K ATP channels are special gates in our cells that help regulate electrical activity, and when they are adjusted, they can reduce nerve excitability and calm pain signals. Opioidergic pathways involve natural pain-relieving chemicals in the body, like endorphins, which bind to receptors in the brain and spinal cord to block pain signals. Meanwhile, serotonergic pathways deal with serotonin, a chemical that helps stabilize mood and can also influence pain perception; when these pathways are enhanced, they can improve pain relief. By influencing these three systems, pain signals are dampened, making us feel more comfortable and less sensitive to pain. Melissa oil reduces swelling (edema), which is achieved by reducing the release of mediators that increase vascular permeability and blood flow. Its primary constituents (geranial and neral, collectively known as citral) reduce the sensitivity of primary nociceptors, which are sensory neurons that detect and transmit pain signals. It is a partial agonist of potential transient receptor type 1 (TRPV1). TRPV1 is a protein that plays a role in pain and inflammation. Citral also inhibits inducible nitric oxide synthase (iNOS) and nuclear factor kappa B (NF-kB). These are key proteins involved in the inflammatory response.

9. Antiparasitic Activity

Molecular or Cellular Target/Pathway: Parasite lifecycle and viability.

Mechanisms: The oil disrupts the lifecycle of these parasites, leading to reduced viability. Melissa essential oil demonstrates activity against *T. brucei* (African sleeping sickness) and *Acanthamoeba* sp. African sleeping sickness is spread to humans through the bite of infected tsetse flies. Once inside the body, the parasite can affect the brain and nervous system, leading to symptoms like fever, headaches, and fatigue. As the disease progresses, it can cause severe sleepiness during the day and difficulty sleeping at night, which is why it's called "sleeping sickness." If not treated, it can be very serious and even life-threatening. Early detection and treatment are crucial for recovery. Acanthamoeba parasites are responsible for several parasitic-related human illnesses including keratitis, encephalitis, acanthamoebiasis, rhinosinusitis, and disseminated infections.

10. Cardiovascular Effects

Molecular or Cellular Target/Pathway: Heart rate regulation.

Mechanisms: Inhalation of melissa oil reduces mean arterial pressure and heart rate in acute coronary syndrome settings. The oil's calming effects can alleviate stress responses in cardiovascular conditions, suggesting its role in overall heart health.

MYRRH

Myrrh essential oil, derived from the resin of the *Commiphora myrrha* (*C. molmol* may be treated as a synonym, but may also be used as a separate species classification) tree, has long been celebrated for its profound therapeutic properties, particularly in traditional medicine. This powerful oil exhibits impressive anticancer activity by inducing apoptosis in several cancer cell types, including breast, liver, and colon cancer cells, through the activation of caspase-dependent pathways—an effect attributed to its key compound, curzerene, which demonstrates selective cytotoxicity toward abnormal cells. In addition to its anticancer properties, myrrh essential oil displays potent anti-inflammatory effects by inhibiting the 5-lipoxygenase enzyme and regulating the NF-κB pathway, potentially alleviating conditions such as gingivitis and gastritis. Furthermore, its robust antimicrobial activity disrupts bacterial cell membranes, providing significant benefits for oral hygiene and wound healing. With additional roles in antiparasitic treatment, cardiovascular protection against oxidative stress, and antioxidant effects that safeguard cellular integrity, myrrh essential oil is a multifaceted ancient remedy worthy of modern usage to enhance overall health and wellness.

SUGGESTED THERAPEUTIC USE:

Cancer: Breast cancer, skin cancer, liver cancer, colon cancer

Inflammatory Conditions: Gingivitis, gastritis, general inflammatory conditions

Infections: Bacterial infections, fungal infections, oral infections, wound infections

Parasitic Infections: Trichomoniasis, fascioliasis

Cardiovascular Protection: Cardiac injury, hypertension, ischemia/reperfusion injury

Organ and Tissue Protection: Oxidative stress-related conditions (cancer, IBD, diabetes, neurodegenerative conditions, cardiovascular disease, liver or kidney disease, metabolic syndrome, fibromyalgia, chronic fatigue syndrome, skin aging), aging, cellular damage, UV damage

Wounds: Wounds, surgical wounds, skin lesions

Oral Health: Gingivitis and other gum diseases

Neurological Disorders: Oral inflammation-related pain, potential systemic inflammatory responses

Insect Control: Mosquito larvae infestations

MOLECULAR AND CELLULAR TARGETS, BIOLOGICAL PATHWAYS, AND MECHANISMS:

1. Anticancer Activity

Molecular or Cellular Target/Pathway: Caspase-dependent apoptosis.

Mechanisms: Myrrh essential oil induces apoptosis in various cancer cell types, including breast, skin, liver, and colon cancer cells. The oil activates caspase-3 and caspase-9 in cancer cells, leading to programmed cell death. Key compounds such as curzerene are cytotoxic to cancer cells while demonstrating reduced toxicity toward normal cells. This selective toxicity highlights its potential as a therapeutic agent.

2. Anti-inflammatory Effects

Molecular or Cellular Target/Pathway: 5-LOX, NF-kB, proinflammatory cytokine production.

Mechanisms: Myrrh oil inhibits the 5-LOX enzyme, reducing the production of leukotrienes involved in inflammation. Myrrh oil regulates the gene expression of inflammatory markers through the inhibition of NF-κB activation. By modulating these inflammatory pathways, myrrh oil potentially alleviates conditions associated with general inflammatory states.

3. Antimicrobial Activity

Molecular or Cellular Target/Pathway: Pathogen cell structure and wall/membrane integrity.

Mechanisms: Myrrh essential oil demonstrates significant activity against various pathogens. The oil disrupts bacterial membrane integrity, leading to cell death and inhibiting biofilm formation. The antimicrobial properties facilitate the reduction of bacterial counts in oral hygiene applications and contribute to wound-healing processes.

4. Antiparasitic Effects

Molecular or Cellular Target/Pathway: Parasite metabolic pathways.

Mechanisms: Myrrh shows efficacy against parasites such as *T. vaginalis* and *Fasciola* species. *T. vaginalis* is a parasite that causes a common sexually transmitted infection called trichomoniasis. This infection can affect both men and women, but often people don't experience any symptoms, which means they can unknowingly spread it. When symptoms do occur, women might notice a foul-smelling vaginal discharge, itching, and burning, while men might have irritation or discharge from the penis. Fasciola species, specifically *F. hepatica* and *F. gigantica*, cause a parasitic infection called fascioliasis. This happens when people eat raw water plants, like watercress, that are contaminated with the parasite's larvae. Once inside, these larvae mature into liver flukes, which then settle in the liver and bile ducts. This can lead to symptoms ranging from fever, nausea, and abdominal pain in the early stages, to more serious problems like liver and bile duct inflammation, and blockages, in later stages. By reducing the egg count and symptoms of fascioliasis, myrrh oil exhibits broad-spectrum antiparasitic activity, potentially benefiting clinical treatment of parasitic infections.

5. Cardiovascular Protection

Molecular or Cellular Target/Pathway: Cardiac proteins and enzymes associated with heart damage, Nrf2, and heme oxygenase-1 pathways.

Mechanisms: Oral administration has been shown to prevent hypertensive responses and alleviate cardiac injury, offering cardioprotective effects, particularly in scenarios of ischemia/reperfusion injury. The oil protects against indices of cardiac injury, including Creatine Kinase-MB (CK-MB), lactate dehydrogenase (LDH), and cardiac troponin T (cTnT). CK-MB, LDH, and cTnT are substances released into your bloodstream when your heart muscle is damaged. These markers, especially cTnT, help doctors detect and assess heart damage, allowing for prompt diagnosis and treatment. Myrrh oil modulates the expression of these antioxidant genes, enhancing cellular defenses against oxidative stress. HO-1 plays a significant role in cardiovascular health, primarily through its protective effects against oxidative stress and inflammation.

6. Antioxidant Effects

Molecular or Cellular Target/Pathway: ROS, oxidative stress.

Mechanisms: Myrrh oil acts as an antioxidant by neutralizing singlet oxygen—a highly reactive form of oxygen that can significantly damage biological molecules, including collagen and elastin, resulting in premature skin aging—and preventing lipid peroxidation, implying its role in protecting skin and systemic health from UV damage. The oil protects plasmid DNA and cellular components from oxidative damage caused by ROS.

7. Wound Healing

Molecular or Cellular Target/Pathway: Wound pathogens.

Mechanisms: Myrrh oil, when combined with other oils, enhances wound healing and microbial protection in lesions and surgical sites. By aiding in microbial control and inflammation reduction, myrrh plays a role in effective wound care protocols.

8. Oral Health (Gingivitis)

Molecular or Cellular Target/Pathway: Inflammatory mediators (TNFa, PGE2, IL-2, iNOS), pathogen inhibition.

Mechanisms: Myrrh oil inhibits oral pathogens associated with gum inflammation and suppresses inflammatory mediators (TNFa, PGE2, IL-2, iNOS).

9. Larvicidal Activity

Molecular or Cellular Target/Pathway: Insect neural pathways.

Mechanisms: Myrrh oil has documented efficacy against larval forms of mosquitoes, indicating potential insect-repellent properties. The exact mechanisms are less defined but likely involve disrupting normal insect development and behavior.

MYRTLE

Myrtle (*Myrtus communis*) essential oil, with its diverse chemotypes, demonstrates a broad spectrum of therapeutic potential. Although it has several different chemotypes, red myrtle (CT 1,8-cineole) and green myrtle (CT alpha-pinene) are the most common commercially available. Characterized by varying concentrations of key constituents like 1,8-cineole, alpha-pinene, and linalool, this oil exhibits promising antitumor, antimicrobial, anti-inflammatory, and hypoglycemic effects. Research reveals its ability to combat drug-resistant pathogens, alleviate inflammatory responses, and regulate blood glucose levels. From inhibiting microbial growth to potentially mitigating cancer cell proliferation and improving wound healing, myrtle oil's applications are extensive. Notably, its effects on oxytocin levels suggest potential benefits for emotional well-being and age-related muscle health. This multifaceted action underscores the complex interplay of myrtle's constituents and their impact on various biological systems.

SUGGESTED THERAPEUTIC USE:

Inflammatory Conditions: Respiratory inflammation (asthma, allergic rhinitis, bronchitis, COPD, pneumonitis, cystic fibrosis, interstitial lung disease, chronic sinusitis, sarcoidosis), chronic inflammatory conditions (rheumatoid arthritis, IBD, psoriasis, COPD, asthma, lupus, eczema, gout, chronic sinusitis, fibromyalgia, chronic fatigue syndrome, periodontitis)

GI Issues: IBD, hemorrhoids, gastric ulcers, intestinal spasms

Metabolic Disorders: Type 2 diabetes, obesity, high cholesterol

Cancer: Colorectal, breast, prostate, ovarian

Respiratory Conditions: Asthma, allergic rhinitis, bronchitis, COPD, pneumonitis, cystic fibrosis, interstitial lung disease, chronic sinusitis, sarcoidosis, congestion, coughs

Infections: Bacterial, fungal, parasitic, tuberculosis, GI infections (*H. pylori*), respiratory infections, viral

Gynecological Conditions: HPV infections

Wounds: Canker sores (aphthous stomatitis), general wounds

Neurological Issues: Cognitive decline, memory

Mental Health Issues: Emotional balance, social interaction

Organ and Tissue Protection: Oxidative stress-related conditions (cancer, IBD, diabetes, neurodegenerative conditions, cardiovascular disease, liver or kidney disease, metabolic syndrome, fibromyalgia, chronic fatigue syndrome, skin aging), cellular damage, aging

MOLECULAR AND CELLULAR TARGETS, BIOLOGICAL PATHWAYS, AND MECHANISMS:

1. Antitumor Activity (Colon Cancer)

Molecular or Cellular Target/Pathway: Cancer cell proliferation, cancer cell proapoptotic pathways.

Mechanisms: Myrtle essential oil induces apoptosis and inhibits cancer cell growth through modulation of signaling pathways.

2. Antimicrobial Activity (Bacteria, Fungi, Protozoa)

Molecular or Cellular Target/Pathway: Pathogen cell walls/membranes, parasitic reproductive cycles.

Mechanisms: Myrtle oil disrupts pathogen cell membrane integrity, inhibits cell wall synthesis, and interferes with metabolic processes, to prevent growth and trigger pathogen death. The oil also enhances antibiotic efficacy through resensitization of drug-resistant pathogens.

3. Anti-inflammatory Activity

Molecular or Cellular Target/Pathway: Leukocyte migration, TNF-α, IL-6, iNOS.

Mechanisms: Myrtle supports a healthy inflammatory response by several mechanisms, including inhibition of proinflammatory cytokine release, reduction of leukocyte chemotaxis to damaged tissue, and suppression of nitric oxide synthase.

4. Hypoglycemic and Lipid-Lowering Effects

Molecular or Cellular Target/Pathway: Alpha-glucosidase, glycolysis, glycogenesis, lipid metabolism.

Mechanisms: Myrtle aids metabolic function by inhibiting carbohydrate breakdown (inhibits alpha-glucosidase), enhancing glucose utilization, modulating lipid synthesis and breakdown, and improving insulin sensitivity. It also affects glycogenesis, which is the process by which glucose molecules are combined to form glycogen, which is stored in the liver and muscles for energy use later. This process helps regulate glucose levels in the body by converting excess glucose into glycogen for storage. It ensures that when blood sugar levels are high, such as after eating, the body can lower those levels by storing the surplus glucose. Later, when blood sugar levels drop, glycogen can be broken down back into glucose (through glycogenolysis) and released into the bloodstream, helping to maintain stable energy levels. Lastly, myrtle improves glycolysis—the metabolic pathway that converts glucose into pyruvate, producing a small amount of energy in the form of ATP, as well as NADH, which can be used in other cellular processes. This process is important because it quickly turns sugar into a form of energy we can use right away, especially when there's not enough oxygen around. It is the first step in energy production for our body. Once glycolysis has done its job, the body can use what it created to produce even more energy through other steps, like the citric acid cycle. Glycolysis also plays a role in making other important substances our bodies need, such as NADH (a molecule key to cellular processes used to produce ATP), ATP, and intermediates that are transformed into amino acids, fatty acids, and nucleotides (the building blocks of DNA and RNA).

5. Antioxidant and Genoprotective Effects

Molecular or Cellular Target/Pathway: Free radical scavenging, DNA protection, endogenous antioxidant activity (SOD, CAT, GPx).

Mechanisms: Myrtle oil protects cells, tissues, and organs by neutralizing ROS, protecting against oxidative DNA damage, and enhancing endogenous antioxidant enzymes (SOD, CAT, GPx).

6. Bronchodilation and Vasodilation

Molecular or Cellular Target/Pathway: Calcium channels in smooth muscle.

Mechanisms: The blockage of calcium influx in airway muscles by myrtle oil results in smooth muscle relaxation and increased airflow/blood flow. This is likely due to its 1,8-cineole content, which is known to block voltage-gated L-type calcium channels in the airways. Blocking certain calcium channels in the muscles of the trachea (the main airway) can help people breathe better. When these channels are blocked, it prevents too much calcium from entering the muscle cells, causing the muscles to relax and the airways to widen. This makes it easier for air to flow in and out of the lungs, especially for those who may have asthma or other conditions that make their airways tight and sensitive. In addition to relaxing the muscles, it can also help reduce the amount of mucus produced in the airways, which can further improve breathing. Overall, this action helps create clearer and more open airways, making breathing easier.

7. Cognitive Effects

Molecular or Cellular Target/Pathway: AChE.

Mechanisms: By inhibiting AChE activity, myrtle increases acetylcholine levels in the synaptic cleft. Elevated acetylcholine can enhance synaptic transmission, which may promote neuronal activity and survival in certain contexts, especially in cases where enhanced cholinergic signaling supports neuronal health. Interestingly, researchers reported that both 1,8-cineole and alpha-pinene were stronger inhibitors of AChE than myrtle oil.

8. Immunomodulatory Effects During Parasitic Infection

Molecular or Cellular Target/Pathway: Innate and adaptive immune responses.

Mechanisms: Myrtle oil has activity against various parasites, including *T. gondii* (toxoplasmosis), *P. falciparum* (malaria), and *T. vaginalis* (trichomoniasis). It helps regulate immune function by enhancing immune cell function, modulating cytokine production, and reducing parasite burden. The research also suggests that myrtle contains constituents that can disrupt the viability and growth of parasites, leading to their elimination. The varying effectiveness at different concentrations and time points may indicate that the oil's mechanism of action involves a gradual disruption of cellular processes or a need for the active components to reach a certain threshold within the parasite cells to be effective.

9. Hemorrhoid Relief

Molecular or Cellular Target/Pathway: TNF-α, IL-6, neutrophil migration, pain pathways, vaso-activity.

Mechanisms: Clinical research shows that myrtle oil helps relieve and heal hemorrhoids. The research concludes this mechanism is due to its anti-inflammatory (reduces proinflammatory cytokines: TNF-α and IL-6; and suppresses the migration of neutrophils to inflamed areas), analgesic and anesthetizing effects (help alleviate pain), and vasoactive properties. While EOs don't contain ions in the traditional sense (like sodium or potassium ions), they do contain vasoactive constituents that can modulate blood vessel tone. This includes 1,8-cineole, linalool, alpha-pinene, beta-pinene, and myrtenyl acetate. These constituents can help contract veins, contributing to its therapeutic effects on hemorrhoids. Hemorrhoids are essentially swollen veins in the anal or rectal area. When veins contract, they become narrower. This narrowing can help reduce the swelling and engorgement of the hemorrhoidal veins, which in turn alleviates the symptoms like pain, bleeding, and itching.

10. Canker Sore (Aphthous Ulcer) Healing

Molecular or Cellular Target/Pathway: Likely antimicrobial, analgesic, antiseptic, anti-inflammatory, and tissue regeneration properties.

Mechanisms: Multiple clinical studies demonstrate that the topical application of myrtle oil speeds the healing of canker sores. While the exact mechanism for this activity is not fully understood, it is likely due to a combination of its antibacterial, antifungal, antiviral, analgesic, antiseptic, anti-inflammatory, and wound-healing properties.

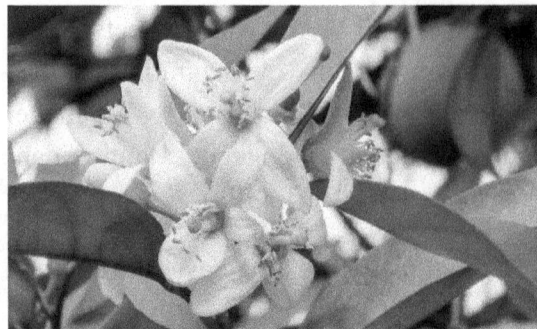

NEROLI

Neroli essential oil, primarily extracted from the blossoms of the *Citrus aurantium* tree, is cherished for its exquisite fragrance and a plethora of therapeutic benefits that promote emotional and physical well-being. Renowned for its anxiolytic and sedative effects, neroli oil balances the GABAergic system, enhancing neurotransmission to alleviate anxiety and muscle tension—particularly beneficial for individuals during labor or postoperative recovery. The oil also offers significant cardiovascular advantages by stimulating vasodilation, aiding blood pressure regulation and reducing stress levels. Furthermore, neroli essential oil possesses anti-inflammatory properties that can soothe postoperative discomfort and chronic inflammatory conditions, while its potent antimicrobial activity combats various pathogens and supports wound healing. Additionally, neroli's antioxidant effects protect cells from oxidative damage, and its ability to modulate hormones, particularly oxytocin, enhances emotional health and social behaviors. With a range of benefits extending to gastrointestinal health and improved sleep quality, neroli essential oil shines as a holistic solution for enhancing overall health and wellness.

SUGGESTED THERAPEUTIC USE:

Mental health Concerns: Anxiety, muscle tension

Cardiovascular Issues: High blood pressure, stress-related cardiovascular issues, poor endothelial function

Inflammatory Conditions: Postoperative inflammation, chronic inflammatory conditions (rheumatoid arthritis, IBD, psoriasis, COPD, asthma, lupus, eczema, gout, chronic sinusitis, fibromyalgia, chronic fatigue syndrome, periodontitis)

Infections: Bacterial infections (e.g., *E. coli*, *S. aureus*, *H. pylori*), fungal infections (e.g., *C. albicans*), wound infections, oral infections, gastrointestinal infections

Organ and Tissue Protection: Oxidative stress-related conditions (cancer, IBD, diabetes, neurodegenerative conditions, cardiovascular disease, liver or kidney disease, metabolic syndrome, fibromyalgia, chronic fatigue syndrome, skin aging), cellular damage, aging

Hormonal Balance: Menopausal symptoms, anxiety related to hormonal fluctuations, emotional health issues

Neuroprotection: Stress-related neurological conditions, potential seizure disorders, convulsive activity

Gastrointestinal Disorders: *H. pylori* infection, gastric ulcers, digestive disorders

Sexual Health: Low libido, postmenopausal sexual discomfort, vaginal dryness

Poor Sleep: Insomnia, sleep disturbances during pregnancy and menopause

MOLECULAR AND CELLULAR TARGETS, BIOLOGICAL PATHWAYS, AND MECHANISMS:

1. Anxiolytic and Sedative Effects

Molecular or Cellular Target/Pathway: GABAergic system (GABA A-BZD receptor).

Mechanisms: Neroli oil has been shown to balance the GABAergic system, which regulates anxiety, muscle tension, and convulsive activity. The oil enhances neurotransmission of GABA, leading to anxiolytic effects and reduction of muscle tension. The anti-anxiety, anticonvulsant, muscle relaxation, and sleep-inducing effects of neroli possibly involve the GABA(A)-Benzodiazepine (GABA A-BZD) receptor complex site. The GABA A-BZD receptor complex is like a master switch in the brain that controls how excitable our brain cells are. When GABA, the brain's natural calming chemical, binds to this switch, it tells the brain cells to slow down. Benzodiazepines, a type of medication, can also bind to this switch and amplify GABA's effect. This slowing down of brain activity helps to reduce anxiety, prevent seizures, induce sleep, and relax muscles.

2. Cardiovascular Benefits

Molecular or Cellular Target/Pathway: NO and soluble guanylate cyclase pathway, and calcium ion modulation.

Mechanisms: Inhalation of neroli oil has been associated with lower cortisol and systolic blood pressure, particularly beneficial for those in stress or high blood pressure situations, such as coronary care patients. Neroli oil stimulates vasodilation, influencing vascular tone and endothelial function. sGC is an important enzyme in the cardiovascular system that is closely associated with NO. Nitric oxide is a signaling molecule produced by endothelial cells that line blood vessels. One of its functions is to promote vasodilation. When NO is released, it binds to and activates soluble guanylate cyclase. Once it's activated, it takes a molecule called GTP and changes it into another molecule called cGMP. Acting as a messenger inside the cells, elevated cGMP sends signals that cause the smooth muscle cells in the walls of blood vessels to relax. When these muscle cells relax, the blood vessels widen (this is called vasodilation). As a result, the blood can flow more freely through the vessels, which helps lower blood pressure and improves the delivery of oxygen and nutrients to different organs in the body. sGC is a target for therapies aimed at treating conditions such as high blood pressure, heart failure, and angina. Neroli also prevents calcium entry into endothelial cells, promoting vasorelaxation and thereby potentially reducing blood pressure.

3. Anti-inflammatory Properties

Molecular or Cellular Target/Pathway: 5-LOX, COX-2, iNOS, and prostaglandin and cytokine synthesis.

Mechanisms: Neroli oil presents mild inhibition of the 5-LOX enzyme. Its constituents modulate the production of inflammatory mediators like IL-6, TNF-α, and prostaglandins through inhibition of COX-2 and nitric oxide production. These properties contribute to reduced inflammation in various conditions, such as in postoperative recovery and chronic inflammatory conditions.

4. Antimicrobial Activity

Molecular or Cellular Target/Pathway: Pathogen membrane integrity.

Mechanisms: Neroli essential oil inhibits the growth of various bacteria and synergizes fluconazole and amphotericin B against *C. albicans*.

5. Antioxidant Effects

Molecular or Cellular Target/Pathway: ROS, oxidative stress.

Mechanisms: Neroli oil acts as a potent antioxidant, capable of scavenging free radicals. This antioxidant capacity helps in protecting cells from oxidative damage, promoting overall health and reducing the risk of chronic diseases.

6. Hormonal Regulation

Molecular or Cellular Target/Pathway: Endocrine function.

Mechanisms: Research suggests that applying neroli to the forearms and inhaling its aroma can improve sexual function in women and reduce menopausal and premenstrual syndrome symptoms. One study observed trends toward improved serum estrogen concentrations. These effects suggest neroli may positively influence hormone balance.

7. Neuroprotective Effects

Molecular or Cellular Target/Pathway: Calcium channel modulation, neurotransmitter release.

Mechanisms: Neroli oil's constituents may protect neuronal function by modulating calcium influx. Calcium influx plays a critical role in the release of neurotransmitters at synapses, facilitating communication between neurons. When a nerve cell gets a signal, it creates a tiny electrical charge that travels down its length. This charge opens doors in the cell's membrane, allowing calcium from the outside to rush in. This influx of calcium acts like a trigger, causing tiny packages called vesicles, which contain chemicals known as neurotransmitters, to merge with the cell's membrane and release their contents into a gap (the synaptic cleft) between nerve cells. The neurotransmitters then travel across this gap to connect with the next nerve cell, helping carry the signal along. Afterward, the body has ways to clear out the neurotransmitters to reset the system for the next signal. Its calming effects and ability to reduce convulsions can contribute to overall neurological health, making it beneficial in both stress-related conditions and potential seizure disorders.

8. Gastrointestinal Health

Molecular or Cellular Target/Pathway: *H. pylori*.

Mechanisms: Neroli oil has been shown to inhibit the growth of *H. pylori* in very low concentrations. The reduction of *H. pylori* may decrease the risk of gastric ulcers and other gastrointestinal disorders, promoting better digestive health.

9. Sexual Health

Molecular or Cellular Target/Pathway: Hormone balance.

Mechanisms: By influencing hormonal levels and increasing blood flow, neroli essential oil can enhance libido and sexual satisfaction in women. Particularly beneficial for postmenopausal and breastfeeding women, the oil can alleviate symptoms associated with hormonal decline, such as vaginal dryness and discomfort during intercourse.

10. Sleep Quality Improvement

Molecular or Cellular Target/Pathway: GABAergic system.

Mechanisms: Inhalation of neroli essential oil can improve sleep quality in individuals, particularly during pregnancy and menopause. The relaxing properties of the oil can help modulate sleep patterns and improve overall restfulness. Activation of the GABAergic system and sedation produced by linalool, linalyl acetate, and limonene interaction with these receptors is the likely mechanism.

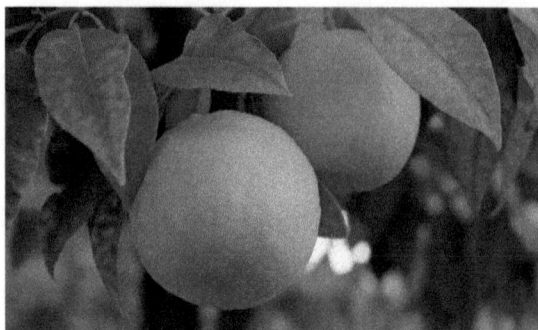

ORANGE (SWEET/WILD)

Orange essential oil, derived from the peel of the *Citrus sinensis* fruit, is a vibrant and versatile oil celebrated for its uplifting aroma and an impressive array of health benefits. Its principal component, d-limonene, has garnered attention for its potential anticancer activity, demonstrating the ability to trigger apoptosis and inhibit tumor growth in various cancer cell lines, including those associated with colon and triple-negative breast cancer. Alongside its cancer-fighting capabilities, orange essential oil exhibits powerful anti-inflammatory effects by significantly inhibiting the 5-lipoxygenase enzyme, thereby reducing proinflammatory cytokines and alleviating conditions such as arthritis. The oil's antioxidant properties act as potent free radical scavengers, protecting cells from oxidative damage, while its mood-enhancing effects and ability to relieve anxiety, particularly through HPA axis regulation, make it invaluable in emotional well-being. Additionally, orange oil possesses notable antimicrobial properties against a variety of bacteria and fungi, supports gastrointestinal health by positively influencing gut microbiota, and promotes skin healing. With its range of therapeutic applications, orange essential oil is a simple but powerful natural remedy for enhancing overall health and wellness.

SUGGESTED THERAPEUTIC USE:

Cancer: Colon cancer, breast cancer, triple-negative breast cancer

Inflammatory Conditions: Arthritis, inflammatory disorders, pain

Organ and Tissue Protection: Oxidative stress-related conditions (cancer, IBD, diabetes, neurodegenerative conditions, cardiovascular disease, liver or kidney disease, metabolic syndrome, fibromyalgia, chronic fatigue syndrome, skin aging), cellular damage, aging

Mental Health Conditions: Anxiety, stress, labor stress, mood disorders

Infections: Bacterial infections, fungal infections, wound infections, biofilm-related infections (chronic wound infections, catheter-associated UTIs, heart valve infections, chronic sinusitis, cystic fibrosis, dental plaque, periodontal disease, osteomyelitis, implant-associated infections, ventilator-associated pneumonia, conjunctivitis, dialysis-associated infections)

Weight Management Metabolic Disorders: Obesity, high cholesterol, metabolic syndrome

Gastrointestinal Disorders: Gut dysbiosis, imbalanced gut microbiota

Sleep Problems: Insomnia, sleep disturbances, postoperative sleep disturbances, childhood sleep disturbances

Insect Control: Mosquito larvae infestations

Wounds and Skin Issues: Skin infections, wounds, slow-healing wounds

MOLECULAR AND CELLULAR TARGETS, BIOLOGICAL PATHWAYS, AND MECHANISMS:

1. Anticancer Activity

Molecular or Cellular Target/Pathway: MAPK and JAK/STAT signaling pathways, angiogenesis, cancer cell proapoptotic pathways.

Mechanisms: Orange essential oil, particularly its main constituent d-limonene, can trigger apoptosis in cancer cells, including those in colon and breast cancer lines. *In vitro* studies show that orange oil can reduce tumor growth by inhibiting angiogenesis (the formation of new blood vessels) and suppressing migration of cancer cells. The oil modulates MAPK and JAK/STAT signaling pathways, inhibiting proliferation and inducing cell death in cancer cells, particularly in triple-negative breast cancer.

2. Anti-inflammatory Effects

Molecular or Cellular Target/Pathway: 5-LOX, proinflammatory cytokines (IL-6, TNF-α).

Mechanisms: Orange oil exhibits a strong inhibitory effect on the 5-LOX enzyme involved in the inflammatory response. The oil reduces the production of proinflammatory cytokines (like IL-6 and TNF-α), thereby decreasing inflammation and pain. The inhibition of inflammatory pathways has implications for conditions such as arthritis and other inflammatory disorders.

3. Antioxidant Activity

Molecular or Cellular Target/Pathway: RO, oxidative stress.

Mechanisms: Orange essential oil serves as a free radical scavenger, mitigating oxidative stress. Studies demonstrate significant antioxidant activity in various assays (e.g., DPPH, β-carotene), offering protective effects against cellular damage.

4. Mood Enhancement and Anxiety Relief

Molecular or Cellular Target/Pathway: NO levels, HPA axis regulation.

Mechanisms: Inhalation or topical application has shown effectiveness in alleviating anxiety, improving mood, and enhancing emotional well-being, especially noted in settings like labor or stressful situations. Orange essential oil affects neurotransmission in the brain, notably modulating nitric oxide (NO) levels, which plays a role in anxiety regulation. NO acts as a neuromodulator in the brain, influencing the release and activity of various neurotransmitters, including serotonin, dopamine, and norepinephrine. These neurotransmitters are crucial for regulating mood and emotional states. Nitric oxide is involved in neuroplasticity, a process that underlies learning, memory, and emotional responses. It modulates synaptic plasticity, particularly in areas of the brain like the hippocampus and amygdala, which are critical for emotional regulation. Research suggests that NO plays a role in the brain's response to stress. Chronic stress can alter NO signaling, which may in turn affect mood and contribute to conditions such as anxiety and depression. It also reduces stress and anxiety responses, potentially balancing the hypothalamic-pituitary-adrenal axis.

5. Antimicrobial Activity

Molecular or Cellular Target/Pathway: Pathogen cell wall/membrane disruption.

Mechanisms: Orange essential oil exhibits substantial activity against various bacteria and fungi. The oil affects bacterial homeostasis and can inhibit the growth of pathogens either through direct toxicity or by disrupting cell membrane function.

6. Weight Management and Metabolic Effects

Molecular or Cellular Target/Pathway: PPAR-γ and acetyl-CoA carboxylase.

Mechanisms: Orange oil influences lipid metabolism by modulating the expression of enzymes involved in fat storage and energy regulation. Studies indicate that oral administration of orange oil can reduce body weight gain and lower cholesterol levels in high-fat diet settings, suggesting potential utility for managing metabolic syndrome. PPAR-γ and ACC are important for body weight management and energy metabolism. PPAR-γ is a protein that helps regulate fat storage and insulin sensitivity, which are crucial for controlling appetite and blood sugar levels. When it is activated, it encourages the body to store fat and use it more efficiently. Acetyl-CoA Carboxylase, on the other hand, is an enzyme that helps produce fat from the carbohydrates we eat, playing a key role in fat metabolism. By influencing how our body processes fat and manages energy, both PPAR-γ and ACC can affect weight gain or loss, making them important targets in the study of obesity and metabolic health.

7. Gastrointestinal Health

Molecular or Cellular Target/Pathway: Gut microbiome.

Mechanisms: Limonene, a key component, has been observed to positively influence gut microbiota composition, increasing beneficial bacteria like Bacteroidetes and *Lactobacillus*. This modulation may improve gut health and immune responses. Bacteroidetes are a group of beneficial bacteria found in our gut that help with digestion, especially breaking down complex foods like fiber into energy. They support gut health by keeping harmful bacteria in check and helping the immune system function properly. Having a good amount of Bacteroidetes in our gut can also lower the risk of obesity and diabetes, making them an important part of a healthy microbiome. *Lactobacillus* is another group of friendly bacteria that are great for gut health. They help digest food and produce beneficial substances like lactic acid, which can keep harmful bacteria away. Lactobacillus also supports the immune system and is often found in yogurt and other fermented foods, which can improve digestion and may help prevent issues like diarrhea. Overall, these bacteria are important for maintaining a balanced and healthy gut.

8. Calming and Sleep Improvement

Molecular or Cellular Target/Pathway: NO neurotransmission, effector memory T cells, CD45+CD11b+ cells.

Mechanisms: The calming effects observed are attributed to its influence on neuroendocrine responses and stress levels, leading to improvements in sleep duration and anxiety. Nitric oxide (NO) is a small molecule in the body that helps with communication between cells, including those in the brain. It plays a role in regulating blood flow, signaling, and inflammation. When NO levels are balanced, they can support healthy brain function, which is important for mood and sleep. However, if NO levels are too high or too low, it can disrupt this balance, potentially leading to increased anxiety, poor mood, and sleep problems. Essentially, keeping NO at optimal levels is important for maintaining emotional stability and good sleep patterns. Effector memory T cells are a type of immune cell that helps the body fight infections, but recent research suggests they might also play a role in mental health, including anxiety. When the body is inflamed or stressed, these cells can release substances that might affect the brain and its ability to regulate mood. Additionally, since there's a strong link between gut health and mental well-being, the immune system, including T cells, could influence gut bacteria, which in turn may affect our feelings of anxiety. While the exact connection between these immune cells and anxiety is still being studied, there's growing interest in how our immune system might impact our mental health. CD45+CD11b+ cells are a type of immune cell found in the meningeal compartment, which is the protective layer around the brain and spinal cord. These cells can become overactive during stress or inflammation, potentially contributing to problems like anxiety and sleep disturbances. Reducing the number of these immune cells in this area could help calm inflammation and create a more balanced environment in the brain. This may lead to better sleep quality and a reduction in anxiety, making it easier for people to relax and feel more at ease. Essentially, by lowering the number of these particular immune cells, we could help our brains function better in terms of sleep and emotional well-being. Orange oil inhalation can promote better sleep quality and reduce disturbances in children and postoperative patients.

9. Insecticidal Properties

Molecular or Cellular Target/Pathway: Direct insect toxicity.

Mechanisms: Orange essential oil possesses larvicidal activity against mosquito species.

10. Skin and Wound Healing

Molecular or Cellular Target/Pathway: Skin and wound pathogens.

Mechanisms: Orange oil shows promise in promoting skin health by enhancing healing and antimicrobial activity against pathogens that cause skin infections.

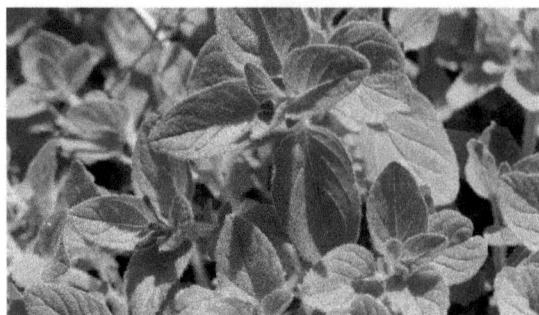

OREGANO

Oregano essential oil, derived from the leaves of the *Origanum vulgare* plant, is a potent oil celebrated for its diverse therapeutic properties and robust pharmacological benefits. Rich in carvacrol and moderate levels of thymol, oregano oil exhibits remarkable anticancer activity, demonstrating the ability to trigger apoptosis in cancer cells and inhibit angiogenesis, thereby hindering tumor growth in various malignancies, including glioblastomas and colon adenocarcinomas. Beyond its anticancer potential, oregano essential oil is renowned for its powerful antimicrobial effects, effectively combating resistant bacterial strains and fungi by disrupting cell membrane integrity and enhancing antibiotic efficacy. Additionally, it offers impressive anti-inflammatory benefits by inhibiting key enzymes, aiding in wound healing, and alleviating inflammation-related ailments. The oil's antioxidant activity provides further advantages, protecting cells from oxidative damage and supporting overall health. With its multifaceted applications—from promoting gastrointestinal health to reinforcing cardiovascular function—oregano essential oil stands out as an exceptional natural remedy, contributing to a holistic approach to wellness and disease prevention.

<u>**SUGGESTED THERAPEUTIC USE**</u>:

Cancer: Glioblastoma, colon adenocarcinoma, cancer cell proliferation, angiogenesis-dependent tumors

Infections: Bacterial infections, fungal infections, antibiotic-resistant infections, biofilm-related infections (chronic wound infections, catheter-associated UTIs, heart valve infections, chronic sinusitis, cystic fibrosis, dental plaque, periodontal disease, osteomyelitis, implant-associated infections, ventilator-associated pneumonia, conjunctivitis, dialysis-associated infections)

Inflammatory Conditions: Inflammation-related conditions

Organ and Tissue Protection: Oxidative stress-related conditions (cancer, IBD, diabetes, neurodegenerative conditions, cardiovascular disease, liver or kidney disease, metabolic syndrome, fibromyalgia, chronic fatigue syndrome, skin aging), cellular damage, aging

Gastrointestinal Infections: *H. pylori* infection, *Entamoeba* infections, gastrointestinal infections

Neuroprotection: Neurodegenerative diseases, cognitive decline, memory impairment

Cardiovascular and Metabolic Health: High oxidized LDL cholesterol, cardiovascular diseases, metabolic syndrome

Insect Control and Repellency: Mosquito infestations, agricultural pests

Parasitic Infections: Parasitic infections (e.g., *Entamoeba*), anisakiasis, giardiasis,

African sleeping sickness, leishmaniasis, American trypanosomiasis

Wound Healing and Skin Restoration: Wounds, scarring, skin regeneration

MOLECULAR AND CELLULAR TARGETS, BIOLOGICAL PATHWAYS, AND MECHANISMS:

1. Anticancer Activity

Molecular or Cellular Target/Pathway: Oncogenes, apoptosis regulators, tumor angiogenesis, tumorigenic pathways, and mitochondrial-mediated proapoptotic pathways.

Mechanisms: In preclinical studies, oregano essential oil has demonstrated efficacy against numerous cancer cell types, including glioblastomas and colon adenocarcinomas, leading to reduced cell viability. Oregano oil components, particularly carvacrol, have been shown to trigger apoptosis in cancerous cells and inhibit angiogenesis (the formation of new blood vessels necessary for tumor growth). Carvacrol has been linked to inhibiting myoblast cell growth after oncogene activation, suggesting it can counteract cancer cell proliferation. When our cells grow and divide normally, they help build and repair tissues. However, sometimes mutations in certain genes (known as oncogenes) can cause cells to grow uncontrollably, which leads to cancer. Inhibiting myoblast cells, which are responsible for muscle growth; after these oncogenes have been activated, the growth of these potentially harmful cells will slow down or stop. This is important because if these myoblasts are prevented from expanding too rapidly, it may help reduce the chances of cancer spreading or developing, especially in tissues where these cells are prominent (skeletal muscle, cardiac muscle, intestines, blood vessels, respiratory tract). It has been noted that oregano oil induces cancer cell death through mitochondrial pathways, affecting gene expression related to cell survival.

2. Antimicrobial Effects

Molecular or Cellular Target/Pathway: Pathogen cell wall and membrane integrity, pathogen mitochondrial homeostasis.

Mechanisms: Oregano oil exhibits strong antimicrobial effects, particularly against various bacteria and fungi. Its phytochemicals (e.g., carvacrol) enhance the activity of antibiotics and potentially reduce pathogenic adherence and biofilm formation, offering alternatives for treating infections. The oil impacts cell membrane integrity, inhibiting the growth of multiple antibiotic-resistant pathogens through enzymatic and structural targets within the bacteria.

3. Anti-inflammatory Properties

Molecular or Cellular Target/Pathway: COX, LOX, oxidative stress.

Mechanisms: Oregano` essential oils significantly inhibit COX and LOX, reducing proinflammatory mediator production. By moderating oxidative stress and inflammation, oregano oil may facilitate faster wound healing and protect against inflammation-related conditions.

4. Antioxidant Activity

Molecular or Cellular Target/Pathway: ROS, oxidative stress.

Mechanisms: Oregano essential oil demonstrates potent antioxidant effects, scavenging free radicals and reducing oxidative stress in cells. Phenols, including compounds like carvacrol, demonstrate higher antioxidant activity primarily due to their chemical structure, specifically the phenolic ring. The hydroxyl (-OH) groups attached to the phenolic ring can donate hydrogen atoms to free radicals, neutralizing them and preventing them from causing oxidative damage to cells. When a phenolic compound donates a hydrogen atom, it forms a relatively stable phenoxyl radical. This radical can delocalize the unpaired electron over the aromatic system, making it less reactive compared to the original free radical, thus limiting further oxidative damage. Some phenolic compounds can chelate (bind) metal ions, such as iron and copper, that catalyze the production of free radicals. Phenolic compounds can also influence the activity of certain enzymes involved in oxidative stress responses, potentially enhancing the body's ability to counteract oxidative damage. Through various assays (e.g., DPPH, ABTS), it is established that oregano oil can protect cellular components from oxidative damage and improve enzymatic antioxidant responses.

5. Gastrointestinal Health

Molecular or Cellular Target/Pathway: Pathogen inhibition and toxicity, pathogen toxin production.

Mechanisms: Oregano oil shows effectiveness against gastrointestinal pathogens and parasites, such as *H. pylori* and *Entamoeba* species. Oregano's phytochemicals prevent toxin formation in pathogens and reduce infection severity, indicating its role in gastrointestinal health and function. Pathogens such as bacteria, fungi, and viruses can produce toxins to harm their host and defend themselves against attacks, like when the body uses antimicrobial agents to fight infections. These toxins are harmful substances created by the pathogens to weaken the immune system or damage cells. For example, some bacteria release proteins (exotoxins) and lipopolysaccharides (endotoxins) that can destroy the cells in our body or disrupt normal bodily functions. Fungi might produce toxins called mycotoxins, which can be harmful if inhaled or consumed. Viruses, while not producing toxins in the traditional sense, can cause cell damage directly as they replicate within host cells, leading to inflammation and symptoms of illness. When antimicrobial agents try to eliminate these pathogens, they might respond by producing more harmful toxins to survive, essentially fighting back to protect themselves and continue infecting their host. This makes treating infections challenging, as the pathogens adapt and evolve in response to the treatment.

6. Neuroprotective Effects

Molecular or Cellular Target/Pathway: AChE, neuronal oxidative stress.

Mechanisms: Inhibition of the enzyme AChE by oregano essential oil helps enhance acetylcholine levels, which are crucial for memory and cognitive function. Research indicates that oregano oil can protect against oxidative stress in neuronal cells, supporting its potential use against neurodegenerative diseases.

7. Cardiovascular Health

Molecular or Cellular Target/Pathway: LDL oxidation.

Mechanisms: The essential oil modulates inflammatory pathways and lipid profiles, lowering cardiovascular disease risk and improving metabolic health outcomes. Oregano oil prevents LDL oxidation, a key factor in cardiovascular diseases. LDL is often referred to as "bad" cholesterol because high levels can contribute to heart disease. However, it's not just the presence of LDL that matters; it's the oxidation of LDL that plays a more significant role in the disease process. When LDL particles become oxidized, they undergo chemical changes that make them more harmful. Oxidized LDL (oxLDL) triggers an inflammatory response in the blood vessels. This response attracts immune cells, such as macrophages, which attempt to clear the oxidized particles. However, when these immune cells consume too much oxidized LDL, they can become overwhelmed and transform into foam cells—macrophages engorged with lipids, which accumulate in the artery walls. This buildup can lead to the formation of plaques, narrowing and hardening the arteries, a condition known as atherosclerosis. As plaques grow, they can restrict blood flow and even rupture, leading to heart attacks or strokes. So, while LDL itself can be a risk factor, it's the oxidation of LDL that sets off a cascade of harmful events that significantly increases the risk of heart disease. Oregano's significant antioxidant activity, ability to suppress lipid peroxidation, anti-inflammatory properties, and potential to regulate endogenous antioxidant activity likely all contribute to the prevention of LDL oxidation.

8. Insecticidal and Repellent Properties

Molecular or Cellular Target/Pathway: Insect nervous systems.

Mechanisms: Oregano oil demonstrates lethal and repellent properties against various insect pests, including mosquitoes and agricultural pests. Its compounds disrupt normal insect behavior and development, providing an eco-friendly alternative to chemical pesticides.

9. Antiparasitic Activity

Molecular or Cellular Target/Pathway: Parasitic cell structures.

Mechanisms: Oregano essential oil exhibits significant activity against several parasites, disrupting their cell membranes and impacting their viability and replication. The essential oil induces oxidative stress and membrane damage in parasitic cells, leading to their death.

10. Wound Healing and Skin Health

Molecular or Cellular Target/Pathway: Modulation of inflammation (MVP-1, VCAM-1, ICAM-1, IP-10, I-TAC, and MIG), tissue remodeling (collagen I, collagen III, EGFR, MMP-1, PAI-1, TIMP-1, and TIMP-2), and the immunomodulatory biomarker M-CSF, and gene expression and signaling related to tissue remodeling.

Mechanisms: Oregano oil modulates inflammatory and tissue remodeling pathways, promoting faster recovery and potentially reducing scarring. Wound healing and skin health involve several key processes and markers that help the body repair itself. Modulation of inflammation refers to how the body controls swelling and redness at the injury site, using signaling molecules like monocyte chemoattractant protein-1 (MVP-1), vascular cell adhesion molecule 1 (VCAM-1), intracellular adhesion molecule 1 (ICAM-1), interferon-γ-inducible protein 10 (IP-10), interferon inducible T cell alpha chemoattractant (I-TAC), and monokine induced by gamma interferon (MIG), which help attract immune cells to fight infection and start healing. Next, tissue remodeling involves rebuilding the skin, with important proteins like collagen I and III providing structure, supported by epidermal growth factor (EGFR) which promotes cell growth, and enzymes like MMP-1 that break down old tissue to make way for new skin. Plasminogen activator inhibitor-1 (PAI-1), tissue inhibitor of metalloproteinases-1 (TIMP-1), and tissue inhibitor of metalloproteinases-2 (TIMP-2) also play a role by regulating these processes and preventing excessive tissue breakdown. Additionally, the molecule macrophage colony-stimulating factor (M-CSF) helps coordinate immune responses and supports the growth of important cells that aid healing. Finally, gene expression and signaling in tissue remodeling refer to how cells communicate and activate specific genes to produce these proteins and enzymes, ensuring the wound heals

properly and the skin remains healthy. Together, these markers and processes orchestrate a complex yet effective response that allows the skin to recover and maintain its integrity.

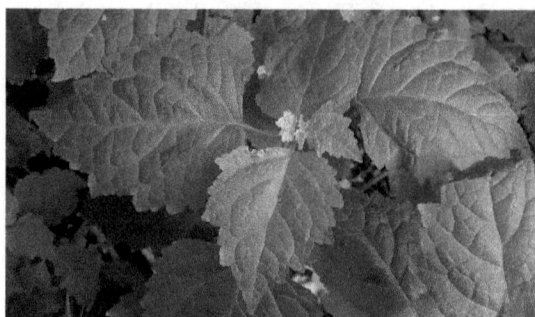

PATCHOULI

Patchouli essential oil, derived from the leaves of the *Pogostemon cablin* plant, is a richly aromatic oil renowned for its extensive therapeutic properties and biological activities. Notably, two of its main constituents, patchoulol and pogostone (Chinese patchouli or patchouli extracted from stems), have demonstrated significant anticancer effects by inhibiting cell proliferation and inducing apoptosis in various cancer cell lines, including colorectal and cervical cancers. In addition to its potential against cancer, patchouli oil offers remarkable anti-inflammatory properties, effectively modulating nitric oxide levels and regulating proinflammatory cytokines to alleviate both acute and chronic inflammation. The oil is also a potent antioxidant, scavenging reactive oxygen species and enhancing the body's enzymatic defenses against oxidative stress. Furthermore, it exhibits strong antimicrobial activity against a range of pathogens, including bacteria and fungi, while promoting immune modulation and gastrointestinal health by favorably altering gut microbiota. With anxiolytic and sedative effects, patchouli oil may support mental health, in addition to providing natural pest control and skin protection benefits. Given its multifaceted array of health-promoting properties, patchouli essential oil is a valuable traditional remedy for both physical well-being and emotional balance.

SUGGESTED THERAPEUTIC USE:

Cancer: Colorectal cancer, cervical cancer, tumor growth

Inflammatory Conditions: Acute inflammation, chronic inflammatory conditions (rheumatoid arthritis, IBD, psoriasis, COPD, asthma, lupus, eczema, gout, chronic sinusitis, fibromyalgia, chronic fatigue syndrome, periodontitis), conditions with excessive leukotriene production (asthma, seasonal allergies, COPD, eosinophilic esophagitis, psoriasis, allergic conjunctivitis, lupus, cystic fibrosis, gout)

Organ and Tissue Protection: Oxidative stress-related conditions (cancer, IBD, diabetes, neurodegenerative conditions, cardiovascular disease, liver or kidney disease, metabolic syndrome, fibromyalgia, chronic fatigue syndrome, skin aging), cellular damage

Infections: Bacterial infections, fungal infections, biofilm-related infections (chronic wound infections, catheter-associated UTIs, heart valve infections, chronic sinusitis, cystic fibrosis, dental plaque, periodontal disease, osteomyelitis, implant-associated infections, ventilator-associated pneumonia, conjunctivitis, dialysis-associated infections), skin odor, food safety

Immune Modulation: Autoimmune disorders, inflammatory immune responses

Gastrointestinal Issues: Gut dysbiosis, leaky gut, imbalanced gut microbiota (dysbiosis)

Mental Health Issues: Anxiety, depression, mental health therapies

Insect Control: Mosquito infestations, insect pests

Skin Protection and Wound Healing: UV-induced skin damage, photoaging, wounds, slow-healing wounds

Longevity and Health Span: Aging, stress resistance, metabolic resilience

MOLECULAR AND CELLULAR TARGETS, BIOLOGICAL PATHWAYS, AND MECHANISMS:

1. Anticancer Activity

Molecular or Cellular Target/Pathway: Cancer cell signaling.

Mechanisms: Patchouli oil's constituents, especially patchoulol and pogostone influence cellular signaling pathways related to apoptosis and cell survival, promoting the programmed death of aberrant cells. Patchouli oil has shown efficacy in decreasing cell proliferation and inducing apoptosis in cancer cells, particularly human colorectal and cervical cancer cells.

2. Anti-Inflammatory Properties

Molecular or Cellular Target/Pathway: 5-LOX, iNOS, proinflammatory cytokines.

Mechanisms: Patchouli oil moderately inhibits the 5-LOX enzyme, which plays a critical role in inflammatory processes. By controlling the production of NO and reducing leukocyte recruitment, patchouli oil can effectively diminish acute and chronic inflammation. It also modulates the release of proinflammatory cytokines (e.g., IL-6, IL-10) and promotes immunomodulatory effects, enhancing the activity of leukocytes and macrophages.

3. Antioxidant Activity

Molecular or Cellular Target/Pathway: ROS, oxidative stress, endogenous antioxidant production.

Mechanisms: Patchouli essential oil demonstrates strong antioxidant activity by scavenging free radicals. The oil activates key antioxidant enzymes such as GPx, SOD, and CAT, aiding in oxidative stress reduction and cellular protection from damage.

4. Antimicrobial Effects

Molecular or Cellular Target/Pathway: Pathogen cell wall/membrane integrity, biofilm formation, direct toxicity.

Mechanisms: Patchouli oil inhibits the growth of various bacteria and fungi. It has been shown to slow the formation of biofilms by harmful bacteria, creating potential therapeutic benefits against biofilm-associated infections. The oil exhibits direct toxicity against various pathogens, including those responsible for skin odor and those affecting food safety, which suggests its utility in food preservation and personal hygiene.

5. Immune Modulation

Molecular or Cellular Target/Pathway: Leukocyte and macrophage activity, proinflammatory cytokine production.

Mechanisms: Patchouli oil modulates the activity of leukocytes and macrophages, key immune cells involved in inflammation and response to infections. Pogostone specifically shows immunomodulatory activity by regulating inflammatory cytokines and suppressing T cell responses, which may be relevant for managing autoimmune disorders.

6. Gastrointestinal Health

Molecular or Cellular Target/Pathway: Gut microbiome, Paneth and goblet cells, and tight junctions.

Mechanisms: Patchouli oil positively alters the gut microbiome by increasing beneficial bacteria, promoting gut health. Research indicates that patchouli oil improves gut microbiome composition, increases the abundance of beneficial bacteria, and enhances the intestinal epithelial barrier. It enhances the production of Paneth and goblet cells important for maintaining mucosal defense and tight junction integrity, reducing the risk of leaky gut.

7. Anxiolytic and Sedative Effects

Molecular or Cellular Target/Pathway: Dopamine.

Mechanisms: Animal studies indicate its mood-balancing properties can help alleviate anxiety and depressive-like behaviors, thus potentially serving as an adjunct in mental health therapies. Inhalation of patchouli oil influences neurotransmitter levels, particularly increasing dopamine in the brain.

8. Insecticidal Activity

Molecular or Cellular Target/Pathway: Insect nervous systems.

Mechanisms: Patchouli oil exhibits insecticidal effects against various pests, including mosquitoes and other insects. The oil can kill pests or disrupt their reproduction and feeding behaviors, suggesting its utility as a natural pest control agent.

9. Skin Protectant

Molecular or Cellular Target/Pathway: ROS, cutaneous antioxidant defenses.

Mechanisms: Patchouli oil enhances the activity of antioxidant enzymes—superoxide dismutase (SOD), catalase (CAT), and glutathione peroxidase (GPx)—to protect the skin from UV-induced damage. These enzymes neutralize ROS, reducing oxidative stress and protecting the skin from UV-induced damage. By improving antioxidant defenses, it helps maintain skin integrity and promotes healing processes.

10. Longevity and Health Span in Aging Models

Molecular or Cellular Target/Pathway: JNK-1/DAF-16 pathway

Mechanisms: Patchouli oil has been shown to activate longevity-associated pathways in model organisms, suggesting effects linked to aging and stress resistance. The JNK-1/DAF-16 pathway is a biological system in our bodies that plays a significant role in helping organisms live longer and healthier lives. When this pathway is activated, it triggers a series of events that help cells adapt to stress, repair damage, and maintain more youthful function. Essentially, JNK-1 is like a messenger that signals the need for a response to stress, and when it does its job, it activates DAF-16, which is a key player in promoting stress resistance, healthy metabolism (glucose and lipid metabolism genes, insulin signaling), and longevity. Together, they help cells function better and protect against age-related diseases, leading to a longer and healthier life. By activating these pathways, patchouli essential oil may promote increased lifespan and health span, enhancing metabolic resilience and reducing oxidative stress.

PEPPERMINT

Peppermint (*Mentha piperita*) essential oil, extracted from the leaves of the plant, is a versatile and invigorating oil renowned for its numerous health benefits and therapeutic properties. Its potent anticancer activity has been demonstrated across various cancer cell types, including lung carcinoma, leukemia, and stomach cancer, primarily through the induction of apoptosis and antioxidant action that reduces oxidative stress. In addition to its anticancer effects, peppermint oil exhibits remarkable anti-inflammatory properties by inhibiting key inflammatory mediators and modulating cytokine signaling pathways, contributing to symptom relief in conditions like irritable bowel syndrome (IBS). The oil's antimicrobial prowess is evident as it disrupts bacterial cell membranes and inhibits biofilm formation, effectively combating a range of pathogenic bacteria and viruses, including herpes simplex virus. Furthermore, peppermint oil supports respiratory health by inducing bronchodilation and alleviating airway spasms, while also enhancing cognitive functions through its interaction with neurotransmitter systems. With its diverse applications—from pain relief and skin health promotion to immune modulation and insect-repellent properties—peppermint essential oil stands out as a powerful ally in holistic wellness.

SUGGESTED THERAPEUTIC USE:

Cancer: Lung carcinoma, leukemia, stomach cancer

Inflammatory Conditions: General inflammation, inflammatory conditions

Infections: Bacterial infections, biofilm-related infections (chronic wound infections, catheter-associated UTIs, heart valve infections, chronic sinusitis, cystic fibrosis, dental plaque, periodontal disease, osteomyelitis, implant-associated infections, ventilator-associated pneumonia, conjunctivitis, dialysis-associated infections)

Gastrointestinal Issues: Irritable bowel syndrome (IBS), gastrointestinal spasms, gut dysbiosis

Viral Infections: Herpes simplex virus (HSV-1, HSV-2) infections

Respiratory Issues: Respiratory distress, airway spasms

Mood and Cognitive Disorders: Cognitive decline, memory impairment, mood enhancement

Wounds: Wounds, slow-healing wounds, skin infections

Pain: Muscle pain, muscle spasms, neuropathic pain

Bone and Joint Conditions: Osteopenia, osteoporosis, Paget's diseases, rheumatoid arthritis, osteoarthritis, gout, juvenile idiopathic arthritis

Insect Control and Repellency: Insect infestations, pest control

Immune Dysregulation: Imbalanced immune responses

Physical Performance: Endurance exercise, athletic performance

MOLECULAR AND CELLULAR TARGETS, BIOLOGICAL PATHWAYS, AND MECHANISMS:

1. Anticancer Activity

Molecular or Cellular Target/Pathway: Cancer cell proapoptotic pathways, oxidative stress.

Mechanisms: Peppermint oil has demonstrated activity against different cancer cell types, including lung carcinoma, leukemia, and stomach cancer cells. Peppermint oil induces apoptosis in cancer cells, which is essential in the mechanism of its anticancer properties. Although mild, its antioxidant properties contribute to reducing oxidative stress, potentially influencing cancer progression.

2. Anti-Inflammatory Properties

Molecular or Cellular Target/Pathway: iNOS, prostaglandin E2, leukocyte recruitment, JAK/STAT signaling.

Mechanisms: Peppermint oil inhibits the production of proinflammatory mediators like nitric oxide and prostaglandin E2. Nitric oxide synthase (NOS) is an enzyme that helps produce a gas called nitric oxide in the body, which plays a crucial role in processes like regulating blood flow and fighting off infections. There are three main isoforms of nitric oxide synthase, which are: 1) eNOS (endothelial nitric oxide synthase)—Primarily found in blood vessels, it is involved in the regulation of vascular tone and blood flow; 2) nNOS (neuronal nitric oxide synthase)—Found in neurons, it is involved in neurotransmission and other neuronal functions; 3) iNOS (inducible nitric oxide synthase)—This isoform is typically expressed in response to inflammatory stimuli and is usually associated with the immune response and pathogenesis. It is usually expressed during inflammatory responses (infections, tissue damage) and produces large amounts of nitric oxide (NO), which can have beneficial roles in combating pathogens but may also contribute to tissue damage and inflammation if produced in excess. By reducing the number of circulating leukocytes and their migration to tissues, peppermint oil decreases inflammation. Peppermint may influence various cytokine signaling pathways related to inflammation, helping to modulate the inflammatory response. JAK/STAT signaling is like a communication system in our cells that helps them respond to various signals, such as those from immune molecules during inflammation. When an inflammatory signal is received, a protein called JAK gets activated, which then turns on another protein called STAT. Once activated, STAT moves into the cell's nucleus and helps switch on genes that can promote inflammation, allowing the body to fight off infections or heal injuries. However, if this signaling is overactive, it can lead to excessive inflammation and contribute to diseases.

3. Antimicrobial Effects

Molecular or Cellular Target/Pathway: Pathogen cell wall/membrane integrity, biofilm formation, efflux pump.

Mechanisms: Peppermint oil effectively inhibits various pathogenic bacteria and demonstrates activity against biofilm formation. The oil disrupts membrane integrity and other cellular functions in bacteria, leading to cell death. Peppermint is antimicrobial through multiple mechanisms of action, including cytoplasmic membrane disruption, increased permeability, potential depolarization, and inhibition of efflux pumps and respiratory activities.

4. Gastrointestinal Health

Molecular or Cellular Target/Pathway: GI smooth muscle, serotonergic activity, GI motility.

Mechanisms: Oral peppermint oil administration significantly alleviates symptoms of irritable bowel syndrome (IBS) and reduces gastrointestinal spasms. Peppermint oil's ability to relax smooth muscle in the GI tract contributes to symptom relief in conditions like IBS. Menthol acts as a calcium channel blocker. This means it interferes with the flow of calcium ions into smooth muscle cells. Calcium ions are essential for muscle contraction. When calcium levels inside the muscle cells decrease, the muscles are unable to contract as strongly, leading to relaxation. By reducing calcium availability in the gut, menthol reduces contractions of the smooth muscle in the gut, thus relieving spasms and cramping. Peppermint also demonstrates serotonergic antagonism, meaning that it can affect the way that serotonin interacts with receptors in the gut. Serotonin is a neurotransmitter that plays a large role in gut function, including gut motility, regulation of gut secretions (mucus, digestive fluids, electrolytes), visceral sensitivity (how sensitive your gut is to pain, bloating, and fullness), and triggering the vomiting reflex. The oil also slows gastrointestinal motility, making it helpful for diarrhea.

5. Antiviral Activity

Molecular or Cellular Target/Pathway: Viral adsorption.

Mechanisms: Peppermint oil exhibits antiviral properties, notably against HSV-1 and HSV-2, even strains resistant to acyclovir. The oil prevents healthy cells from being infected with the virus, showcasing its potential use in treatment protocols for viral infections.

6. Respiratory Health

Molecular or Cellular Target/Pathway: Calcium channels in airway smooth muscle, TRPM8, IL-6/JAK2/STAT3 pathway.

Mechanisms: Peppermint oil can cause bronchodilation, helping to alleviate symptoms of respiratory distress. Similar to its gut effects, menthol can influence calcium channels in airway smooth muscle. By interfering with calcium influx, it reduces the ability of these muscles to contract, leading to bronchodilation. Menthol activates transient receptor potential melastatin 8 (TRPM8) receptors, which are sensitive to cold and are found on sensory nerve fibers in the airways. Activation of TRPM8 can lead to a sensation of coolness and can also contribute to bronchodilation, possibly through reflex mechanisms or by influencing nerve signaling. Menthol, a primary component, aids in reducing airway spasms and improves overall respiratory function. Peppermint oil can affect inflammatory pathways like the IL-6/JAK2/STAT3 pathway, reducing inflammation, and subsequently constriction, in the airways. Menthol can have a direct relaxant effect on airway smooth muscle, independent of nerve signaling. This may involve interactions with other ion channels or intracellular signaling pathways.

7. Cognitive Benefits

Molecular or Cellular Target/Pathway: AChE, GABA(A), nicotinic receptors.

Mechanisms: Peppermint oil is shown to enhance cognitive performance and mood, likely through interaction with neurotransmitter systems. Peppermint oil binds to GABA(A) and nicotinic receptors and inhibits AChE. The oil's ability to inhibit AChE may help in preserving acetylcholine levels, thus supporting memory and cognitive functions. By binding to GABA receptors, components of peppermint oil may produce a similar calming effect, allowing for more focused attention. When acetylcholine binds to nicotinic receptors, it typically leads to increased signaling in the brain. This can improve attention, memory, and cognition. Lastly, by increasing intracellular calcium levels, it is possible that components of peppermint oil may increase signaling in the brain.

8. Skin Health and Wound Healing

Molecular or Cellular Target/Pathway: Fibroblast activity, collagen synthesis, wound pathogens.

Mechanisms: Peppermint oil promotes skin health by enhancing fibroblast activity and collagen synthesis, critical for wound closure. It reduces infection rates in wounds and promotes healing through antimicrobial effects and improved tissue regeneration.

9. Pain Relief and Muscle Relaxation

Molecular or Cellular Target/Pathway: TRP channels.

Mechanisms: The oil is utilized in massage therapies for muscle relaxation and pain management. Peppermint's pain-relieving mechanism is multifaceted, primarily revolving around how its main component, menthol, interacts with our sensory system. The oil targets sensory receptors, interrupts pain signals, relaxes muscles, and lessens inflammation. Peppermint oil provides pain relief through its cooling effect on the skin and muscles. Peppermint, and particularly its main component menthol, exerts its pain-relieving effects through several cellular and molecular targets. A key mechanism involves interactions with TRP channels, which are proteins involved in sensing various stimuli, including temperature and pain. Menthol activates TRPM8, which is primarily responsible for the cooling sensation. This activation can also provide analgesic effects, especially in cases of neuropathic pain. Menthol can also desensitize TRPA1, which is involved in detecting irritants and inflammatory pain. Menthol can also sensitize TRPV3, which may help modulate the pain response, leading to better pain management. Peppermint's ability to affect calcium channels can relax muscles, reduces spasms, and affect nerve signaling.

10. Insecticidal and Repellent Properties

Molecular or Cellular Target/Pathway: Insect nervous systems.

Mechanisms: Peppermint oil has demonstrated significant insecticidal and repellent activity against various pests. The oil disrupts the feeding and reproductive behaviors of insects, suggesting its potential role in pest control.

11. Immunomodulation

Molecular or Cellular Target/Pathway: Leukocyte and macrophage activity, proinflammatory cytokines, gut microbiome.

Mechanisms: Peppermint oil appears to enhance immune responses by modulating the activity of leukocytes and macrophages. It regulates the release of various cytokines, promoting a balanced immune response while reducing excessive inflammation. Peppermint also favorably influences the gut microbiome, increasing Collinsella bacteria and improving the Firmicutes/Bacteroidetes ratio in children. *Collinsella* can influence immune responses through various mechanisms, including interactions with the gut mucosal barrier and the production of metabolites that affect immune cell function. Research suggests that early colonization of *Bifidobacterium* and *Collinsella* populations in children during the first months of life supports healthy childhood development and growth. Firmicutes/Bacteroidetes ratio is considered a health biomarker that is disrupted in obesity, metabolic disorders, and inflammatory conditions. A healthy ratio can improve gut barrier integrity, overall immune cell activity, and the production of short-chain fatty acids (SCFAs).

12. Bone and Joint Health

Molecular or Cellular Target/Pathway: Osteoclasts, synoviocytes.

Mechanism: Menthol, a main ingredient found in peppermint oil, has beneficial effects on our joints and bones. It helps keep bones strong by slowing down the activity of certain cells called osteoclasts, which break down bone tissue. By doing this, menthol can help prevent bones from becoming weak or thin, which can lead to conditions like osteoporosis. Additionally, menthol can help reduce swelling and discomfort in the joints by encouraging the death of overly active cells called synoviocytes. These cells line the joints and can multiply too much when there is inflammation, leading to pain and stiffness. By reducing the number of these cells, menthol can help ease inflammation and make it easier to move. Overall, menthol's ability to support bone strength and reduce joint inflammation makes it a valuable compound for promoting joint health.

13. Athletic and Physical Performance

Molecular or Cellular Target/Pathway: Central nervous system, bronchial smooth muscle tone, oxidative damage.

Mechanism: Several studies have investigated the effects of peppermint essential oil on physical performance, both through inhalation and ingestion. Oral administration of peppermint essential oil improved grip force, standing vertical jump, standing long jump, forced vital capacity, peak inspiratory flow rate, peak expiratory flow rate, and reaction times. Inhalation led to increases in running speed, hand grip strength, and the number of push-ups performed. However, it did not affect skill-related tasks such as basketball free-throw accuracy. Inhalation of peppermint essential oil significantly extended the exhaustion time in rats subjected to weight-bearing swimming training. Menthol can affect athletic performance through mechanisms related to its thermal, ventilatory, analgesic, and arousing properties, influencing the perception of fatigue and the regulation of exercise performance. Peppermint may improve spirometric measurements—tests that help assess lung function by measuring how much air you can breathe in and out—by affecting bronchial smooth muscle tone. Peppermint essential oil has antioxidant and anti-inflammatory properties, which are believed to reduce oxidative stress and inflammation induced by exercise, thus relieving exercise-induced fatigue and improving exercise performance. Peppermint essential oil treatment led to reduced oxidative damage induced by endurance exercise. These studies collectively suggest that peppermint essential oil, whether ingested or inhaled, may have beneficial effects on physical performance metrics.

ROMAN CHAMOMILE

Roman chamomile (*Chamaemelum nobile*, syn. *Anthemis nobilis*) essential oil, extracted from the flowers of the plant, is celebrated for its soothing and therapeutic qualities that benefit both the mind and body. Notably, this gentle oil boasts potent anti-inflammatory properties, achieved through the mild inhibition of key enzymes such as 5-lipoxygenase (5-LOX) and cyclooxygenase-2 (COX-2), thereby reducing the synthesis of inflammatory mediators and effectively alleviating various skin conditions. In addition to its anti-inflammatory effects, Roman chamomile essential oil exhibits impressive anxiolytic and sedative properties, promoting emotional well-being by increasing oxytocin levels and modulating stress responses. Its ability to support neurogenesis and decrease stress hormone levels also suggests potential antidepressant effects. Furthermore, this essential oil offers skin-soothing and healing benefits, assists in digestive relief, and demonstrates antimicrobial actions against both bacteria and parasites, making it a multifaceted remedy with a wide range of applications. From relaxation and pain relief to enhanced skin care, Roman chamomile essential oil is a valuable addition to holistic wellness practices, providing gentle yet effective support for physical and emotional health.

SUGGESTED THERAPEUTIC USE:

Inflammatory Conditions: Inflammatory skin conditions, general inflammation

Mental Health Issues: Anxiety, stress, emotional well-being, social behavior issues, depression, mood disorders, cognitive resilience

Skin Issues and Wounds: Inflammatory skin conditions, wound healing, eczema, psoriasis

Hormone Imbalance: Vaginal atrophy

Sexual Health: Postmenopausal sexual function and desire (libido)

Muscular Issues: Age-related muscle loss

Pathogen Infections: Parasitic infections (e.g., *Haemonchus contortus*), bacterial infections

Life Transitions: Palliative care support

Digestive Issues: Nausea, digestive discomfort

MOLECULAR AND CELLULAR TARGETS, BIOLOGICAL PATHWAYS, AND MECHANISMS:

1. Anti-Inflammatory Activity

Molecular or Cellular Target/Pathway: 5-LOX, COX-2, prostaglandin and leukotriene synthesis.

Mechanisms: Roman chamomile essential oil mildly inhibits this enzyme, which is involved in leukotriene production and inflammation. The oil reduces the production and activity of COX-2, which is upregulated during inflammatory responses, without affecting COX-1, thus potentially sparing beneficial processes mediated by COX-1. By inhibiting COX-2 and 5-LOX, Roman chamomile reduces the synthesis of proinflammatory prostaglandins and leukotrienes, contributing to its anti-inflammatory effects. Roman chamomile can penetrate deeper skin layers to reduce inflammatory skin conditions, evidenced by decreased levels of inflammatory mediators.

2. Anxiolytic and Sedative Properties

Molecular or Cellular Target/Pathway: Likely GABAergic-Benzodiazepine (GABA A-BZD), and central nervous system, nerve growth factor.

Mechanisms: The oil's inhalation during stressful situations, such as coronary angioplasty, suggests its potential to modulate the body's stress response effectively. Many sedative and anxiolytic drugs exert their effects by modulating the GABAergic system, the primary inhibitory neurotransmitter system in the central nervous system. It is plausible that compounds in chamomile essential oil, such as angelic acid esters, may enhance GABA activity through direct GABA binding, increasing its release, or inhibiting its reuptake and degradation. Some compounds in chamomile EO may interact with benzodiazepine-binding sites, contributing to its sedative and anxiolytic properties. Reduced motility in a rodent study indicates a general depressant effect on the central nervous system. This could involve decreasing excitatory neurotransmitters (glutamate), modulating ion channels related to neuronal excitability, or effects on serotonin and dopamine. In animal studies, the essential oil seems to enhance NGF levels, which may play a role in reducing anxiety and improving mood and cognitive functions. NGF is a special protein in our bodies that helps develop and maintain nerve cells, which are crucial for how our brain communicates and processes feelings. When levels of NGF are balanced, it supports healthy brain function, including mood and stress responses. However, if the level of NGF is too high or too low, it can lead to issues with how we feel and handle anxiety. Higher levels of NGF have been linked to increased anxiety and mood disorders because they can affect the connections between nerve cells in the brain, making it harder to cope with stress or negative emotions. NGF plays a role in keeping our mood stable, and any imbalance can lead to feelings of anxiety or sadness.

3. Antidepressant Effects

Molecular or Cellular Target/Pathway: Doublecortin (DCX)-positive cells in the dentate gyrus, neurite outgrowth, neurogenesis, corticosterone.

Mechanisms: Roman chamomile essential oil increases the proliferation of neurons in the hippocampus when used alongside antidepressants like clomipramine. Doublecortin (DCX)-positive cells are a type of young nerve cell found in a part of the brain called the dentate gyrus, which is important for memory and mood regulation. When these cells are forming properly, they help keep our mood stable and may protect against feelings of depression. However, studies have shown that in people who are experiencing depression, the number of these young nerve cells is often lower. This decrease in DCX-positive cells might mean that the brain is having a hard time generating new nerve cells, which can lead to a poorer ability to adapt to stress and emotional difficulties. In simple terms, having fewer DCX-positive cells can signal problems in the brain that are linked to feelings of sadness or depression. The essential oil potentially supports neurogenesis, which is crucial for improving mood and cognitive resilience. Neurite outgrowth and neurogenesis are two important processes in the brain that help create and strengthen connections between nerve cells. Neurite outgrowth refers to the growth of tiny extensions from nerve cells that help them communicate with one another. Neurogenesis is the creation of new nerve cells, especially in areas that affect mood and learning. When we're feeling depressed, these processes can be disrupted, meaning our brain might struggle to form new connections and generate new cells. This can lead to feelings of hopelessness and difficulties in coping with stress. In simple terms, when the brain isn't firing on all cylinders—by making new cells or strengthening existing connections—it can contribute to feelings of depression. Roman chamomile reduces corticosterone, a stress hormone, thereby possibly alleviating depressive symptoms and enhancing overall mood.

4. Skin Soothing and Healing Properties

Molecular or Cellular Target/Pathway: LPS-induced COX-2 mRNA and protein expression.

Mechanisms: The oil influences various inflammatory markers at the cellular level, promoting skin healing in conditions like vaginal atrophy. When applied topically, Roman chamomile soothes inflammatory skin conditions through its bioactive compounds, enhancing healing and comfort. In one study, chamomile decreased LPS-induced COX-2 mRNA and protein expression. This suggests that chamomile works similarly to non-steroidal anti-inflammatory drugs (NSAIDs).

5. Antimicrobial and Antiparasitic Activity

Molecular or Cellular Target/Pathway: Direct toxicity, pathogen cell wall/membrane integrity.

Mechanisms: Roman chamomile has been shown to effectively inhibit the motility of parasites (e.g., *Haemonchus contortus*) and exhibits effectiveness against certain bacteria. The oil's compounds, particularly isobutyl angelate, inhibit important life processes of the target organisms, thereby demonstrating antifungal and antiparasitic properties. The oil disrupts membrane integrity and other cellular functions in bacteria, leading to cell death.

6. Digestive Relief

Molecular or Cellular Target/Pathway: Possibly olfactory system and vagus nerve.

Mechanisms: Aromatherapy with a blend containing Roman chamomile, anise, fennel, and peppermint has demonstrated efficacy in reducing nausea associated with various hospice and palliative care treatment. The olfactory system has strong connections to the limbic system, which in turn interacts with areas of the brainstem that are also connected to the vagus nerve. This means that olfactory stimuli can influence the vagus nerve indirectly through these interconnected brain regions. Specifically, areas like the amygdala and hypothalamus, which are influenced by olfaction, also have connections to the vagal nuclei. The vagus nerve also significantly influences olfaction (bidirectional relationship). Studies involving vagus nerve stimulation have demonstrated effects on the olfactory bulb, indicating that the vagus nerve can modulate olfactory processing. Stimulating the vagus nerve can help improve olfactory function and reverse olfactory dysfunction. The aroma of Roman chamomile helps modulate the central nervous system's response to nausea, providing symptom relief.

7. Emotional and Social Behavior, Sexual Health

Molecular or Cellular Target/Pathway: Oxytocin.

Mechanism: Oxytocin is a neuropeptide and hormone produced in the hypothalamus and released by the posterior pituitary that influences emotions, mental health, social behaviors, and sexual arousal. Interestingly, research also suggests that declining oxytocin levels during aging contributes to decreases in skeletal muscle mass and strength. Inhaling Roman chamomile significantly increases oxytocin levels in postmenopausal women, suggesting it could potentially improve mental and emotional health, social behaviors, and sexual desire, and may help preserve muscle mass and strength during aging.

ROSE

Rose essential oil, derived from the exquisite petals of the *Rosa damascena* plant, is prized not only for its intoxicating fragrance but also for its remarkable therapeutic benefits. This precious oil exhibits potent anti-inflammatory effects by mildly inhibiting key enzymes such as 5-lipoxygenase (5-LOX) and cyclooxygenase-2 (COX-2), contributing to the reduction of inflammatory markers associated with conditions like dysmenorrhea and muscle spasms. Also revered for its analgesic properties, rose oil activates transient receptor potential vanilloid 1 (TRPV1) receptors to modulate pain perception, offering natural relief. Beyond its physical benefits, rose essential oil nurtures emotional well-being through its anxiolytic and antidepressant effects, as it influences neurotransmitter pathways and elevates oxytocin levels, promoting relaxation and improved mood. Furthermore, its muscle relaxant and neuroprotective activities complement its broad spectrum of uses, while its strong antimicrobial properties make it effective against various pathogens. Enhanced sleep quality and hormonal balance are additional benefits that contribute to the oil's holistic appeal, making rose essential oil a cherished partner in the pursuit of wellness and balance.

SUGGESTED THERAPEUTIC USE:

Inflammatory Conditions: Skin inflammation

Pain: Pain, pain perception

Mental Health Issues: Anxiety, depression, stress, postpartum depression, stress-related neuronal damage

Muscle Relaxant Activity: Tracheal muscle spasms, intestinal spasms, muscle spasms,

Cardiovascular Disorders: High blood pressure

Neuroprotective Effects: Oxidative stress-related neuronal damage, Parkinson's disease-related neuronal damage, stress-related neuronal damage

Infections: Bacterial infections, fungal infections

Hormonal Effects and Female Reproductive Issues: Menstrual symptoms, menopausal symptoms, hormonal imbalances, dysmenorrhea

Sleep Disorders: Insomnia, anxiety-related sleep disturbances

Skin Issues: Hyperpigmentation, skin lightening

MOLECULAR AND CELLULAR TARGETS, BIOLOGICAL PATHWAYS, AND MECHANISMS:

1. Anti-Inflammatory Effects

Molecular or Cellular Target/Pathway: 5-LOX, COX-2, prostaglandin synthesis.

Mechanisms: Rose oil mildly inhibits the 5-LOX enzyme, involved in the inflammatory response, thereby reducing leukotriene production. Similar to 5-LOX, the oil can inhibit COX-2, contributing to the reduction of proinflammatory prostaglandins without affecting COX-1. Inhibition of COX-2 leads to decreased synthesis of inflammatory prostaglandins. By reducing the activity of COX-2 and 5-LOX, rose essential oil decreases levels of inflammation-related mediators, aiding conditions like dysmenorrhea, muscle spasms, and skin inflammation.

2. Analgesic Properties

Molecular or Cellular Target/Pathway: TRPV1.

Mechanisms: Rose oil activates TRPV1 receptors, which play a critical role in pain sensation. TRPV1 receptors are special proteins found in your body, particularly in your nerves. They play a role in sensing pain, heat, and certain chemicals. Activation of TRPV1 can lead to temporary alterations in pain signal processing in the nervous system and reduction of pain perception. Instead of continuously sending strong pain signals, activating TRPV1 can reduce the sensitivity of the nerves. Activating TRPV1 may also stimulate the release of natural pain-relieving substances in your body, called endorphins. These help improve your mood and further reduce the sensation of pain.

3. Anxiolytic and Antidepressant Effects

Molecular or Cellular Target/Pathway: Oxytocin, nerve growth factor, GABA(A).

Mechanisms: Rose oil influences neurotransmitter pathways associated with anxiety and mood regulation, potentially modulating serotonin levels. It has been observed to improve symptoms of anxiety and depression, particularly in postpartum women and those undergoing medical procedures. Inhalation of rose oil has been shown to increase oxytocin levels, which play a role in social behavior and stress reduction. Enhanced growth factors like nerve growth factor (NGF) lead to neuroprotection and potentially counteract stress-related neuronal damage. When you experience stress, it can harm neurons and lead to feelings of anxiety and depression. By increasing NGF levels, you can promote the growth and survival of these nerve cells, helping to repair any damage caused by stress. This not only protects your brain but also improves its ability to function properly, which can lead to better mood regulation and a reduction in anxiety and depression symptoms. Essentially, NGF acts like a protective shield for your brain, making it more resilient and better equipped to handle stress. Rose also interacts with GABA(A) receptors and unknown non-benzodiazepine mechanisms, and its constituents 2-phenethyl alcohol and citronellol are known to possess anxiolytic properties.

4. Muscle Relaxant Activity

Molecular or Cellular Target/Pathway: Possibly NO-cGMP pathway.

Mechanisms: Rose oil has been shown to relax smooth muscles in various applications, including in the trachea and intestines. Specifically, it promotes relaxation in vascular smooth muscle, which can lead to lower blood pressure and improved respiratory function. One of the mechanisms by which rose oil may exert this effect is through the NO-cGMP pathway. In this process, nitric oxide (NO) is released and acts like a key that unlocks the relaxation of blood vessels. Once released, NO diffuses into the smooth muscle cells surrounding the blood vessels and activates an enzyme called guanylate cyclase. This enzyme then produces cyclic guanosine monophosphate (cGMP), which initiates a series of events that relax the smooth muscles in the blood vessels. As a result, this relaxation leads to the widening of blood vessels and an increase in blood flow.

5. Neuroprotective Effects

Molecular or Cellular Target/Pathway: Oxidative stress pathways, direct antioxidant activity, neurotransmitters, BDNF, calcium homeostasis, inflammatory pathways (RIPK1/eIF4E), α-synuclein expression, dopamine, Bcl-2/Bax ratio, cytochrome c and caspase-9 expression, and NF-κB and p38 MAPK pathways.

Mechanisms: Rose oil protects against oxidative stress, particularly in neurons exposed to toxic substances like aluminum and drugs like L-dopa used in Parkinson's disease therapy. This suggests that the oil may counteract metal accumulation-related neural damage. It reduces production of reactive oxygen species (ROS) and prevents lipid peroxidation in neuronal tissues. After a spinal cord injury, there is often inflammation, oxidative stress, and cell death, which can worsen the damage. Rose oil (geraniol) counteracts these harmful processes, protecting the spinal cord. Rose oil may also influence neurotransmitters (e.g., acetylcholine release) and inflammatory pathways (e.g., iNOS, proinflammatory cytokines). Citronellol, the major monoterpene alcohol of rose oil, is an RIPK1 agonist and modulates the RIPK1/eIF4E pathway. RIPK1 is a protein that can trigger inflammation in the brain, while eIF4E is a protein that helps cells make other proteins and can also play a role in increasing inflammation. RIPK1 can activate eIF4E, which then leads to more inflammation in the brain. This inflammation can contribute to problems like depression. So, RIPK1 and eIF4E work together in a way that can cause harmful inflammation in the brain. Some rose oil constituents influence calcium homeostasis and neurotrophic factor pathways (BDNF), which play a role in neuroprotection and neural plasticity. Furthermore, one of its major constituents, geraniol, is highly neuroprotective, especially in Parkinson's disease models. It reduces α-synuclein expression, increases dopamine levels, reduces neuron death, normalizes Bcl-2/Bax ratio and decreases cytochrome c and caspase-9 expression, and modulates NF-κB and p38 MAPK pathways. This means that geraniol can improve the various signs and symptoms associated with conditions like Parkinson's disease by improving behavior, brain chemistry, and the levels of specific proteins in the brain. In Parkinson's disease, a protein called α-synuclein can clump together in the brain, forming harmful aggregates that damage nerve cells. Geraniol helps to prevent this clumping, which can help protect the nerve cells from damage. Dopamine is a chemical messenger in the brain that is critical for movement. In Parkinson's disease, the brain cells that produce dopamine are damaged, leading to reduced dopamine levels and movement problems. Geraniol can help increase the amount of dopamine in the brain, which can help improve movement. When cells undergo apoptosis, they go through a programmed process of self-destruction. In conditions like Parkinson's disease, too much apoptosis can occur, leading to the loss of important nerve cells. Geraniol can help to reduce this excessive cell death, protecting the nerve cells. By normalizing the balance of Bcl-2 and Bax, and by decreasing cytochrome c and caspase-9, geraniol helps to regulate the cell death process and prevent excessive nerve cell loss. NF-κB and p38 MAPK are important signaling pathways in cells that control inflammation and other processes. By modulating these pathways, geraniol helps to regulate inflammation and protect cells from damage.

6. Antimicrobial Activity

Molecular or Cellular Target/Pathway: Pathogen cell wall/membranes.

Mechanisms: Rose oil exhibits strong antimicrobial activity against a variety of pathogens. Compounds in rose oil affect the integrity of microbial cell membranes, leading to pathogen death and inhibition of growth.

7. Hormonal Effects

Molecular or Cellular Target/Pathway: Estrogen.

Mechanisms: Inhalation of rose oil can elevate salivary estrogen levels, which may influence hormonal balance in women. By promoting hormone balance, rose is a promising solution for menstrual symptoms, and hormonal regulation, especially in relation to menopause.

8. Sleep Quality and Stress Reduction

Molecular or Cellular Target/Pathway: HPA axis, GABA.

Mechanisms: Rose oil inhalation has been established to positively impact the stress response and improve sleep quality in various populations. Its benefits are driven by regulating the stress response through the HPA axis. In one study, rose oil produced an anti-anxiety effect similar to diazepam, a common anxiolytic drug. This suggests that rose oil may act on the same biological pathways as diazepam. Diazepam is known to bind to GABA receptors in the brain, which are part of the central nervous system and help to calm activity in the brain. Constituents like citronellol and 2-phenethyl alcohol also regulate the body's response to stress and anxiety. Through its mild sedative effects, rose oil can help alleviate anxiety, facilitate relaxation, and improve sleep quality, making it beneficial for caregivers and patients in high-stress situations.

9. Antimelanogenic Activity

Molecular or Cellular Target/Pathway: Tyrosinase, ROS.

Mechanisms: Rose inhibits the activity of the enzyme responsible for melanin production (tyrosinase), thereby lightening the skin and reducing hyperpigmentation. Inhibiting ROS production helps prevent skin conditions related to abnormal pigmentation.

ROSEMARY

Rosemary (*Rosmarinus officinalis*) essential oil, derived from the aromatic leaves of the plant, is renowned for its versatile medicinal properties that range from enhancing cognitive function to exhibiting powerful anticancer effects. This versatile oil has shown promising potential in inducing apoptosis in cancer cells by modulating the expression of BCL2 and BAX genes, while also protecting DNA and inhibiting tumor proliferation across various cancer types, including ovarian, liver, and breast tumors. Alongside its anticancer capabilities, rosemary essential oil boasts potent anti-inflammatory properties by inhibiting 5-lipoxygenase (5-LOX) and activating the Nrf2 pathway, which enhances the body's antioxidant defenses. Its antimicrobial activity against pathogens such as *E. coli* and *C. albicans* further establishes its role in holistic health. Additionally, the oil promotes cognitive enhancement by inhibiting acetylcholinesterase, thereby elevating acetylcholine levels crucial for memory. With properties that soothe pain and improve circulation, rosemary essential oil is a natural solution no home should be without—combining the power of nature with therapeutic benefits to support overall well-being.

SUGGESTED THERAPEUTIC USE:

Cancer: Ovarian cancer, liver cancer, colon cancer, breast cancer, prostate cancer, DNA damage, cancer cell proliferation

Inflammatory Conditions: General inflammation, chronic diseases, oxidative stress-related inflammation, excessive leukocyte migration

Infections: Bacterial infections, fungal infections, biofilm-related infections,

Insect Control: Insect infestations

Neurological Defects: Memory impairment, cognitive decline, neurodegenerative diseases, neuronal damage

Organ and Tissue Protection: Oxidative stress-related conditions (cancer, IBD, diabetes, neurodegenerative conditions, cardiovascular disease, liver or kidney disease, metabolic syndrome, fibromyalgia, chronic fatigue syndrome, skin aging), cellular damage

Pain: General pain

Circulatory Conditions: Poor circulation, Raynaud's phenomenon

Skin Conditions and Wounds: Eczema, bullous pemphigoid, wounds

Metabolic Disorders: Diabetes, hyperglycemia, high cholesterol, metabolic disorder

MOLECULAR AND CELLULAR TARGETS, BIOLOGICAL PATHWAYS, AND MECHANISMS:

1. Anticancer Effects

Molecular or Cellular Target/Pathway: BCL2 and BAX gene expression, DNA protection, and cancer cell proapoptotic pathways.

Mechanism: Rosemary oil contains compounds that have been shown to act as regulators of cell proliferation, potentially influencing the growth and division of cells in a way that may protect against certain diseases or support overall cellular health. It inhibits tumor formation and proliferation across multiple tissues. It also modulates the expression of BCL2 and BAX genes, promoting apoptosis and weakening cell membranes in cancer cells. BCL2 and BAX are two important genes that play a role in controlling cell life and death. Think of BCL2 as a "helper" gene that tries to keep cells alive, while BAX acts like a "trigger" that can lead to cell death when things go wrong. In cancer, the balance between these two genes can be disturbed. For example, if BCL2 is overly active, it can prevent cancer cells from dying when they should, allowing tumors to grow and spread. On the other hand, if BAX is not working properly, it can also lead to cells living longer than they should. Lastly, rosemary oil protects against DNA damage, contributing to its anticancer effects.

2. Anti-inflammatory Properties

Molecular or Cellular Target/Pathway: 5-LOX, Nrf2, leukocyte activity.

Mechanism: Rosemary contains compounds that can inhibit the enzyme 5-lipoxygenase (5-LOX), which is involved in the inflammatory response, thereby potentially reducing inflammation and offering protective effects against

conditions like cancer and other chronic diseases. Rosemary has been found to activate the Nrf2 pathway, a crucial regulator of antioxidant defense mechanisms, which helps to combat oxidative stress and may contribute to the prevention of various chronic diseases and the promotion of overall health. It may also interact with leukocyte receptors, potentially influencing immune responses. By reducing excessive leukocyte migration and adhesion to damaged tissues, rosemary oil prevents abnormal inflammatory responses.

3. Antimicrobial Effects

Molecular or Cellular Target/Pathway: Pathogen cell walls/membranes, biofilm formation.

Mechanism: Rosemary displays broad-spectrum antimicrobial activity against a variety of pathogens. It works by disrupting microbial cell walls/membranes, inhibiting biofilm formation, and immune stimulation.

4. Cognitive and Neurological Effects

Molecular or Cellular Target/Pathway: AChE, choline levels, Gap43 gene expression, T-type calcium channels, and oxidative stress.

Mechanism: Rosemary oil inhibits acetylcholinesterase (AChE), which can impact neurotransmitter regulation and has potential implications for neurological and cognitive functions. It modulates intracellular acetylcholine, choline, and Gap43 gene expression levels and inhibits T-type calcium channels. T-type calcium channels are special pathways in brain cells that allow calcium ions to enter the cells. The Gap43 gene is important for brain health because it helps produce a protein that plays a key role in the growth and repair of nerve cells. When the Gap43 gene is expressed, it supports the development of new connections between neurons, which is essential for learning, memory, and recovery after brain injuries. Higher levels of Gap43 help brain cells communicate better and adapt to changes, making it vital for neuroprotection and overall brain function. When these channels open, they can trigger various processes that help protect the brain from injury or damage, such as during a stroke or neurodegenerative disease. By regulating calcium influx, T-type channels help neurons communicate better, support cell survival, and even promote healing after injury. In essence, they play a role in making sure brain cells can cope with stress and stress conditions, contributing to overall brain health and protection. Rosemary oil protects neurons from oxidative stress, contributing to cognitive enhancement. Lastly, it reduces neuronal excitability, enhances neuron communication, and improves neuron survival and healing after injury.

5. Antioxidant Activity

Molecular or Cellular Target/Pathway: ROS, MDA, antioxidant activity, endogenous antioxidants (CAT, SOD, GSH).

Mechanism: Rosemary reduces production of reactive oxygen species (ROS), limiting oxidative stress and damage to cells and tissues. It exhibits antioxidant capacity, reducing oxidative stress markers like ROS and MDA. MDA, or malondialdehyde, is a byproduct that forms when cells are under oxidative stress. Oxidative stress occurs when there are too many free radicals, which are unstable molecules that can damage cellular components like DNA, proteins, and lipids. When fats in the cell membranes are oxidized, MDA is produced as a result. Because elevated levels of MDA can indicate increased oxidative damage, it is often used as a marker to evaluate the extent of oxidative stress in the body. The oil also supports endogenous antioxidant activity (CAT, SOD, and GSH), which are essential antioxidant enzymes that help protect cells from oxidative damage by neutralizing reactive oxygen species.

6. Pain Relief and Vasodilation

Molecular or Cellular Target/Pathway: Opioid receptors, serotonin receptors, alpha-1 and alpha-2 adrenergic receptors.

Mechanism: Rosemary oil relieves pain by interacting with both opioid and serotonin receptors. It also stimulates alpha-1 and alpha-2 adrenergic receptors. Alpha-1 and alpha-2 adrenergic receptors are special receptors that respond to certain chemicals, like adrenaline. When it comes to pain, these receptors play different roles. Alpha-1 receptors, when activated, can help increase blood flow to an area, which might lead to inflammation and the sensation of pain. On the other hand, alpha-2 receptors help to control pain by reducing the release of other chemicals that make us feel pain. Think of it like a light switch: turning on the alpha-1 receptors can make the pain experience brighter, while activating the alpha-2 receptors can help dim that pain. By modulating the activity of these two receptors, rosemary oil diminishes pain and how the body responds to it.

7. Alleviation of Raynaud's Phenomenon

Molecular or Cellular Target/Pathway: Thermoregulation, alpha-2 adrenergic receptors, endothelial function.

Mechanism: Rosemary oil promotes localized, and possibly systemic, blood flow by acting on alpha-2 adrenergic receptors. The blood vessels are lined with smooth muscles. When these muscles tighten, the blood vessels narrow, reducing blood flow. When they relax, the blood vessels widen, and blood flows more easily. By regulating alpha-2 adrenergic receptors, rosemary oil causes the smooth muscle to relax, dilating the blood vessels and improving blood flow. In Raynaud's phenomenon, the small blood vessels in the fingers and toes overreact to cold or stress. They suddenly narrow too much, causing the fingers and toes to become pale, cold, and sometimes painful due to reduced blood flow. Rosemary oil helps counteract this by promoting the relaxation of the smooth muscles in the blood vessels, which leads to improved blood flow. This helps to alleviate the symptoms of Raynaud's phenomenon by warming the fingers and toes.

8. Cardiovascular Benefits

Molecular or Cellular Target/Pathway: Adrenergic receptors.

Mechanism: Rosemary essential oil improves circulation by stimulating alpha-1 and alpha-2 adrenergic receptors in the heart, blood vessels, and kidneys. It restores skin barrier function and reduces inflammation through specific signaling pathways.

9. Metabolic Regulation

Molecular or Cellular Target/Pathway: Antioxidant activity, inflammation, antiapoptotic pathways, pro-proliferative activity, glucose uptake and utilization, lipid metabolism.

Mechanism: Rosemary oil improves glycemic control and reduces oxidative stress, often complementing pharmaceutical interventions. It demonstrates high antioxidant activity, effectively scavenging free radicals and reducing oxidative stress in kidney tissue. This is evidenced by a substantial reduction in malondialdehyde (MDA) levels and an increase in total glutathione (GSH) levels after use. Another kidney-protective factor is the ability of rosemary to reduce inflammation, reducing kidney damage caused by diabetes. Additionally, rosemary oil showed antiapoptotic effects, helping to prevent kidney cell death. It simultaneously exhibits pro-proliferative activity, which aids in the repair and regeneration of kidney tissues. Kidney protection is important for glucose control because the kidneys play a crucial role in regulating blood sugar levels and overall metabolic health. When blood sugar levels are high, the kidneys filter out excess glucose, which is then excreted in the urine. However, over time, high blood sugar can damage the kidneys, leading to a condition called diabetic nephropathy, which impairs their ability to function properly. This can create a vicious cycle where kidney damage further affects blood sugar control, making it harder to manage diabetes effectively. Additionally, healthy kidneys help regulate blood pressure and fluid balance, both of which are important for overall health in people with diabetes. Rosemary oil lowers blood glucose levels in diabetic subjects by enhancing glucose uptake and utilization, decreasing glycogen content, increasing glycolytic rate (the speed at which cells break down glucose to produce energy) in liver and muscle cells, and inhibiting glucose production and gluconeogenesis (the process by which the liver produces glucose from non-carbohydrate sources, like proteins and fats) in liver cells. Lastly, the oil regulates lipid metabolism and improves lipid profiles—reducing total cholesterol and triglycerides and increasing HDL cholesterol.

10. Insecticidal Properties

Molecular or Cellular Target/Pathway: Insect nervous systems.

Mechanism: Rosemary disrupts insect acetylcholinesterase (AChE), which insects rely heavily on for neural communication, especially in motor control and signaling within their nervous systems. When AChE is disrupted, acetylcholine remains in the synaptic cleft longer than normal. This leads to sustained stimulation of postsynaptic receptors, resulting in continuous and uncontrolled signaling. Eventually, the disruption of AChE causes a critical failure in the insect's nerve signaling pathways, leading to their death.

SANDALWOOD

Sandalwood (*Santalum album, S. paniculatum, S. spicatum, S. austrocaledonicum*) essential oil, particularly derived from *Santalum album*, exhibits a multifaceted array of therapeutic properties that make it a valuable asset in both clinical and holistic applications. Extensive research has demonstrated its potential in combating various forms of cancer, notably bladder and skin cancers. Through mechanisms such as preventing viral replication, especially against herpes simplex viruses, sandalwood protects against viral assaults. Additionally, sandalwood oil is recognized for its potent anti-inflammatory effects, largely due to its ability to inhibit the 5-LOX enzyme and reduce proinflammatory mediators. Its antiparasitic and antimicrobial activity is notable against pathogens like *Plasmodium falciparum* and *Clostridium perfringens*, while also demonstrating efficacy in treating skin conditions such as acne and psoriasis. Beyond its physical health benefits, sandalwood oil enhances psychological well-being by promoting alertness, balancing autonomic function, and potentially increasing oxytocin levels, which contributes to improved emotional and social health. As a whole, sandalwood essential oil serves as a versatile therapeutic agent with promising implications for various health conditions.

<u>SUGGESTED THERAPEUTIC USE</u>:

Cancer: Bladder cancer, skin cancer, skin papillomas, HPV-related lesions

Viral Infections: HSV-1 infections, HSV-2 infections, drug-resistant herpes infections

Inflammatory Conditions: General inflammation, leukotriene-mediated inflammation, skin inflammation

Antimicrobial Effects: Wound infections, gas gangrene, eumycetoma

Parasitic Infections: Malaria

Neurosensory Conditions and Cognition: Low alertness, impaired memory, cognitive dysfunction

Mental Health Issues: Low oxytocin levels, mental health issues, emotional health issues, social behavior issues, autonomic nervous system imbalance, mood instability

Sexual Function: Low sexual desire

Skin Health Applications: Acne, psoriasis, eczema, warts, dermatological conditions

Pain: Inflammation-related pain, oxidative stress-related pain

Cardiovascular Conditions: Doxorubicin-induced cardiotoxicity, cardiac injury

MOLECULAR AND CELLULAR TARGETS, BIOLOGICAL PATHWAYS, AND MECHANISMS:

1. Anticancer Effects

Molecular or Cellular Target/Pathway: Proapoptotic pathways, UV-induced AP-1 activity.

Mechanism: Sandalwood has demonstrated activity against various forms of cancer, including bladder cancer, skin cancer, and skin papillomas. Its primary constituent, alpha-santalol, has been widely investigated for its activity against skin cancers. The oil induces apoptosis and hinders cell replication in bladder cancer cells. Sandalwood inhibits pre-cancerous cells and reduces UV-induced AP-1 activity, offering protective benefits against skin cancer. It does so by reducing the progression of pre-cancerous cells to skin cancer and inhibiting AP-1, a signaling molecule that triggers skin cancer development. AP-1 is a group of proteins that help control how cells respond to stress and inflammation, including damage from UV rays from the sun. When UV light hits the skin, it can activate AP-1, leading to changes in the skin cells that may promote the development of cancer. By inhibiting AP-1 activity, it's possible to reduce these harmful changes, helping to protect the skin and lower the risk of skin cancer. Lastly, topical application of sandalwood reduces the number and spread of skin papillomas related to the human papillomavirus (HPV), which may be precursors to squamous cell skin cancer. Skin papillomas are benign growths on the skin that can be caused by HPV. When HPV infects the skin, it can lead to the rapid growth of skin cells, resulting in these warts. While most papillomas do not pose any health risks and may not require treatment, certain strains of HPV can be associated with cancers, particularly in other areas of the body, such as the cervix, throat, and anus. However, the papillomas themselves, like common warts or skin tags, are usually considered non-cancerous.

2. Antiviral Activity

Molecular or Cellular Target/Pathway: HSV-1 and HSV-2.

Mechanism: Sandalwood influences multiple targets involved in the replication and virulence of herpes simplex virus (HSV) strains. It prevents replication of HSV-1 and HSV-2, exhibiting noteworthy virucidal activity against both drug-resistant and non-drug-resistant strains. Disrupts HSV viral envelope—a lipid bilayer, or "envelope," that normally protects them from the outside environment. When enveloped viruses lose their envelope, they can no longer infect cells. Decreases viral plaque formation, indicating that new viral particles are not being produced as efficiently in the presence of sandalwood oil.

3. Anti-inflammatory Properties

Molecular or Cellular Target/Pathway: 5-LOX, COX, proinflammatory mediators (cytokines, chemokines, prostaglandins).

Mechanism: Sandalwood oil strongly suppresses the 5-Lipoxygenase (5-LOX) enzyme, which is involved in the conversion of arachidonic acid to leukotrienes, which are potent proinflammatory mediators, thereby reducing inflammation. Sandalwood counters proinflammatory mediators, such as cytokines, chemokines, prostaglandins, and COX enzymes. Specific to the skin, sandalwood oil prevents the formation of various proinflammatory mediators in skin cells, mimicking the effects of ibuprofen.

4. Antiparasitic and Antimicrobial Effects

Molecular or Cellular Target/Pathway: Malaria parasite (*P. falciparum*).

Mechanism: Sandalwood oil demonstrates inhibitory action against the malaria-causing parasite *in vitro*. *P. falciparum* is one of the most dangerous species of malaria-causing parasites. It is primarily transmitted to humans through the bite of infected female *Anopheles* mosquitoes. It can cause severe illness and complications, including cerebral malaria, anemia, and multi-organ failure, making it a leading cause of malaria-related deaths worldwide. Unlike some other malaria species, *P. falciparum* can rapidly multiply in the bloodstream, contributing to its severity and the challenges in treatment and prevention.

Molecular or Cellular Target/Pathway: Wound pathogens.

Mechanism: A combination of sandalwood and myrrh essential oil displays significant antimicrobial effects against wound-associated bacteria. The oil exhibits antibacterial activity against gas gangrene bacteria. Its highest inhibition was observed against *C. perfringens*, particularly when combined with palmarosa oil. Gas gangrene is a potentially deadly condition that occurs when bacteria—usually *C. perfringens* or *C. septicum*—gather in a wound that has no blood supply and produce toxins that release gas, causing tissue death. Additionally, sandalwood oil displays considerable antifungal activity against the fungus causing eumycetoma—*M. mycetomatis*. Eumycetoma is a chronic infectious disease that affects the skin, tissues, and bones, typically resulting in painful swelling and the formation of lumps or nodules. It is caused by certain types of fungi, with *M. mycetomatis* being one of the most common culprits. Often found in tropical and subtropical regions, eumycetoma can occur after the fungus enters the body through cuts or wounds. The disease can be quite difficult to treat, and if left unmanaged, it may lead to serious complications or permanent damage.

5. Neurosensory Effects

Molecular or Cellular Target/Pathway: Brain activity, cognitive function networks.

Mechanism: Inhalation of sandalwood oil increases arousal and attentiveness, as indicated by physiological changes (e.g., pulse rate and blood pressure). This is likely mediated by altering activity in brain regions related to alertness (prefrontal cortex, etc.). Inhaling the oil also enhances the brain's cognitive function networks. Specifically, it results in beta and gamma band activation in the cingulate gyrus of the prefrontal cortex, suggesting improved memory maintenance.

6. Psychological and Emotional Effects

Molecular or Cellular Target/Pathway: Oxytocin.

Mechanism: Inhalation of sandalwood increases salivary oxytocin levels, possibly enhancing mental and emotional health, social behaviors, and sexual desire in postmenopausal women.

7. Adaptogenic Properties

Molecular or Cellular Target/Pathway: Autonomic nervous system function.

Mechanism: Massage with sandalwood balances autonomic function, promoting harmonized states and moods.

8. Skin Health Applications

Molecular or Cellular Target/Pathway: Pathogens associated with skin conditions (e.g., acne, psoriasis, eczema, warts).

Mechanism: Sandalwood shows potential for treating various skin conditions in clinical trials due to its anti-inflammatory and antimicrobial properties.

9. Cardioprotection

Molecular or Cellular Target/Pathway: Endogenous antioxidants (GSH, SOD, CAT), inflammatory mediators (IL-1β, TNF-α, NF-κB), cardiac protection and function.

Mechanism: Sandalwood essential oil provides cardioprotection primarily by enhancing antioxidant defenses and reducing inflammation, thereby limiting cardiac damage caused by doxorubicin (DOX). Administration of the oil with DOX protects against cardiotoxicity by reducing inflammation and oxidative stress in the heart. Specifically, sandalwood enhances the levels of antioxidants such as GSH, SOD, and CAT, and inhibits the intense inflammatory response following cardiac injury affecting inflammatory mediators such as IL-1β and TNF-α, as well as the signaling protein NF-κB. Sandalwood led to lower levels of cardiac injury biomarkers (like creatine phosphokinase and cardiac troponin T), improvements in heart rhythm (ECG changes), and stabilization of blood pressure and heart rate. These improvements indicate that sandalwood helps preserve overall cardiac function affected by DOX.

SPEARMINT

Spearmint (*Mentha spicata*) essential oil, primarily composed of carvone, exhibits a diverse range of therapeutic effects that contribute to its use in both traditional and contemporary medicinal practices. Its efficacy in alleviating nausea/vomiting associated with chemotherapy highlights its importance as a complementary treatment alongside standard antiemetic medications. Spearmint oil's notable calcium channel-blocking properties enhance cardiovascular health by lowering blood pressure and improving lipid profiles. In addition, its potent antimicrobial and antifungal activities suggest vast applications in treating infections caused by bacteria and fungi, including drug-resistant strains. The oil also demonstrates significant anti-inflammatory effects and may alleviate pain associated with osteoarthritis and other conditions. Furthermore, the biologically active compounds in spearmint essential oil show promise in agricultural applications as natural crop protectants. Overall, the multifaceted benefits of spearmint essential oil position it as a valuable resource for enhancing health and combating various ailments.

SUGGESTED THERAPEUTIC USE:

Gastrointestinal Issues: Nausea, vomiting, post-surgical nausea

Cardiovascular Conditions: High blood pressure

Pain: Smooth muscle spasms, pain, nociception, neuropathic pain

Infections: bacterial infections, fungal infections, biofilm-related infections (chronic wound infections, catheter-associated UTIs, heart valve infections, chronic sinusitis, cystic fibrosis, dental plaque, periodontal disease, osteomyelitis, implant-associated infections, ventilator-associated pneumonia, conjunctivitis, dialysis-associated infections)

Inflammatory Conditions: Acute inflammation, chronic inflammatory conditions (rheumatoid arthritis, IBD, psoriasis, COPD, asthma, lupus, eczema, gout, chronic sinusitis, fibromyalgia, chronic fatigue syndrome, periodontitis)

Organ and Tissue Protection: Oxidative stress-related conditions (cancer, IBD, diabetes, neurodegenerative conditions, cardiovascular disease, liver or kidney disease, metabolic syndrome, fibromyalgia, chronic fatigue syndrome, skin aging), cellular damage

Insect Control and Repellency: Insect infestations, insect bites

Hormone Imbalance and Female Reproductive Issues: PCOS

Neurological Defects: Alzheimer's disease, neurological inflammation

Cancer: Cancer cell growth, oxidative stress in cancer

Athletic Performance/Improvement of Lung Function: Enhanced lung function, enhanced athletic/physical performance

MOLECULAR AND CELLULAR TARGETS, BIOLOGICAL PATHWAYS, AND MECHANISMS:

1. Antiemetic Effects (Reduction of Nausea and Vomiting)

Molecular or Cellular Target/Pathway: Likely targets—receptors and pathways involved in the vomiting reflex (CNS and GI), 5-HT3 and dopamine receptors, GI motility and secretion.

Mechanisms: Spearmint, in combination with peppermint, appears to reduce the frequency and severity of nausea and vomiting, potentially by modulating neurotransmitter activity in the brain and affecting gastrointestinal motility. The body has a vomiting reflex to protect itself, which involves complex communication between the brain (CNS) and the gut (GI).

When something irritates the stomach or the brain signals danger (like toxins), it can trigger nausea and vomiting. When chemicals like serotonin bind to 5HT-3 receptors in the gut (often when sick or after eating something bad), signals are sent to the brain to trigger nausea and vomiting. Medications, stress, and motion sickness can activate dopamine receptors and trigger nausea or vomiting. The brain's "vomiting center" processes signals from the gut and other areas (like the inner ear for motion). When it gets signals about potential harm, it commands the body to vomit. Nausea can slow down the movement of the stomach and intestines, which changes how food and fluids are processed. Sometimes, it can also lead to increased secretion of stomach acid or bile, both of which can contribute to the sensations that cause vomiting. Inhalation of spearmint in a blend with other oils also reduced post-surgical nausea, likely through similar mechanisms.

2. Antispasmodic and Cardiovascular Effects

Molecular or Cellular Target/Pathway: Calcium channels and signaling, smooth muscles.

Mechanisms: Carvone, a major component of spearmint oil, acts as a calcium channel blocker, relaxing smooth muscle and potentially reducing blood pressure. This mechanism is similar to that of verapamil.

3. Antimicrobial and Antifungal Activity

Molecular or Cellular Target/Pathway: Pathogen cell walls/membranes, enzymes and proteins essential for microbial survival, microbial energy pathways, biofilm formation.

Mechanisms: Spearmint oil inhibits the growth of various bacteria and fungi, by disrupting cell membrane integrity, interfering with essential metabolic processes involved in pathogen energy production, and inhibiting biofilm formation. Spearmint oil has demonstrated greater efficacy than some antifungal drugs.

4. Anti-inflammatory and Analgesic Effects

Molecular or Cellular Target/Pathway: COX-2, proinflammatory cytokines, nociceptors, arachidonic acid pathway.

Mechanisms: Spearmint oil exhibits anti-inflammatory and antinociceptive activity, possibly by inhibiting COX-2, reducing inflammatory cytokine production, and modulating pain perception. Pain can be reduced by targeting nociceptors, which are the nerve endings in our body that sense harmful stimuli like injury or inflammation. When you hurt yourself, these nociceptors send pain signals to your brain, making you feel discomfort. Certain essential oils, like spearmint, can block these signals or reduce the sensitivity of nociceptors. This can help decrease the amount of pain you feel.

5. Antioxidant Activity

Molecular or Cellular Target/Pathway: ROS, endogenous antioxidants.

Mechanisms: Spearmint oil scavenges free radicals and enhances antioxidant enzyme activity, protecting cells from oxidative damage.

6. Insecticidal and Repellent Activity

Molecular or Cellular Target/Pathway: Insect nervous system receptors, insect reproductive systems.

Mechanisms: Components of spearmint oil act as biocides and repellents against various insects, potentially by disrupting their nervous system function and interfering with their reproductive processes.

7. Effects on Polycystic Ovarian Syndrome (PCOS)

Molecular or Cellular Target/Pathway: Ovarian follicular development, testosterone levels.

Mechanisms: Spearmint oil may reduce testosterone production and restore normal follicular development in PCOS, potentially by modulating steroid hormone synthesis. It notably reduces the number of atretic follicles (degenerated follicles) and ovarian cysts, and increases the number of Graafian follicles (mature follicles). PCOS is characterized by hyperandrogenism, which is an excess of androgens, including testosterone. In PCOS, excess testosterone can disrupt the balance of hormones necessary for the development of ovarian follicles, which contain the eggs. This can lead to the formation of cysts and other symptoms of PCOS. Testosterone can be converted into a more potent androgen called dihydrotestosterone (DHT). DHT may contribute to mitochondrial fission in granulosa cells of PCOS patients. By reducing testosterone levels, treatments like spearmint oil can help restore hormonal balance and improve ovarian function. More recently, researchers have identified that PCOS is strongly associated with metabolic dysregulation. Many women with PCOS struggle with insulin resistance, meaning their bodies don't use insulin effectively. Obesity, particularly abdominal obesity, is frequently observed in PCOS, as is abnormal lipid profiles. These metabolic disturbances are so common in PCOS that they're often classified as metabolic syndrome, indicating that PCOS is more than just a hormonal disorder; it also has significant metabolic components. Carvone, the main component of spearmint oil, helps protect the insulin-producing cells in the pancreas from damage and plays a key role in managing how our body processes carbohydrates. It supports important enzymes that help regulate blood sugar levels—glucokinase, glucose-6-phosphate dehydrogenase (G6PD), glucose-6-phosphatase, and fructose-1,6-bisphosphatase, leading to better liver function and lower levels of certain liver enzymes (AST, ALT, and ALP). By doing this, carvone lowers blood sugar and helps keep insulin working properly, which is crucial for maintaining stable glucose levels in the body. Additionally, it aids in improving fat metabolism by reducing triglycerides and cholesterol levels, which contributes to overall better health and metabolic balance.

8. Neurological Effects

Molecular or Cellular Target/Pathway: AChE, BChE, aryl hydrocarbon receptor (AhR).

Mechanisms: Spearmint oil shows inhibitory activity against AChE and BChE (stronger against AChE than against BChE), which is linked to Alzheimer's disease. Spearmint oil also inhibits the AhR pathway, which is important for immune and other cellular functions. Spearmint oil has also been seen to reduce neuropathic pain. AhR is a protein found in our cells that helps regulate how our body responds to various environmental signals, including pollutants and toxins. It plays a crucial role in brain health by influencing the development and function of neurons (nerve cells) and the overall balance of chemicals in the brain. When activated, AhR can affect the expression of genes involved in inflammation and neurotransmitter production, which are essential for proper brain function. If there is too much activation or dysregulation of AhR, it may contribute to neurological problems, such as mood disorders and neurodegenerative diseases. Essentially, AhR acts like a gatekeeper, helping to maintain a healthy balance in the brain, which is important for our mental and emotional well-being.

9. Anticancer Activity

Molecular or Cellular Target/Pathway: Expression of BRAF, EGFR, KRAS, NFKB1, NFKBIA, NFKB2, PIK3CA, PIK3R, PTEN, TP53, AKT1, AKT2, FOS, and RAF1 genes, cancer cell proapoptotic pathways, COX-2, MAPK, EGFR/p38.

Mechanisms: Spearmint oil exhibits cytotoxicity to various cancer cells and increases expression of antioxidant genes, showing promise in cancer research. This cytotoxicity leads to a decrease in cell viability and triggers apoptosis, a form of programmed cell death. Spearmint oil exhibits selective cytotoxicity toward cancer cells, meaning it effectively kills cancer cells while showing minimal toxicity to healthy cells. Spearmint oil and its constituents can alter the expression of genes involved in cancer development and progression. For instance, specific compounds in spearmint oil have been shown to increase the expression of genes like BRAF, EGFR, KRAS, NFKB1, NFKBIA, NFKB2, PIK3CA, PIK3R, PTEN, and TP53, while decreasing the expression of others, such as AKT1, AKT2, FOS, and RAF1. These genes play critical roles in various cellular processes, including cell growth, proliferation, and survival. Spearmint oil has been shown to inhibit COX-2, an enzyme that plays a significant role in inflammation and cancer. By inhibiting COX-2, spearmint oil can reduce inflammation and interfere with the pro-carcinogenic effects mediated by this enzyme. Spearmint oil can induce apoptosis in cancer cells through the caspase pathway. Spearmint oil may disrupt the interaction between MAPK activity, EGFR/p38, and COX-2, which is known to promote angiogenesis (the formation of new blood vessels) in cancer cells. By interfering with this interaction, spearmint oil can potentially inhibit angiogenesis and suppress cancer growth.

10. Enhanced Athletic Performance

Molecular or Cellular Target/Pathway: Airway dynamics and respiratory function.

Mechanism: Inhalation of spearmint essential oil prior to vigorous exercise significantly enhances lung function—forced expiratory volume 1 (FEV1) and forced vital capacity (FVC), resulting in improved athletic performance. FEV1 and FVC are indicators of better lung function. FEV1 measures how much air you can force out of your lungs in just one second, while FVC tracks the total amount of air you can exhale after taking a deep breath. When athletes have higher levels of FEV1 and FVC, it means their lungs can take in and deliver more oxygen to their muscles more efficiently. This is important during exercise, as muscles need oxygen to perform well and sustain energy levels. Therefore, better lung capacity helps athletes breathe easier, enhances their stamina, and ultimately improves their overall athletic performance.

TANGERINE/MANDARIN

Tangerine (*Citrus reticulata* Blanco, syn. *C. nobilis*) and mandarin (*C. reticulata*) essential oils, rich in the bioactive compound limonene, exhibit a wide range of therapeutic effects, highlighting their potential as a natural remedy for various health concerns. Tangerine is actually a subtype of mandarin and to add to the confusion, there are multiple types of mandarin essential oils commercially available—green, yellow, and red—depending on the maturity of the fruit at the time of distillation. The primary distinguishing factor between tangerine and mandarin is their limonene and gamma-terpinene content. Tangerine usually contains greater than 80% limonene and 5% or less of gamma-terpinene, while mandarin has higher levels of gamma-terpinene ($\geq$ 13%) and a maximum of 80% limonene. Research demonstrates their ability to inhibit inflammation through the suppression of the 5-lipoxygenase enzyme and prevent the formation of harmful aflatoxins produced by molds. Additionally, these oils exhibit significant antimicrobial, antiviral, and antiparasitic properties, making them a promising candidate for controlling infections and pests. Moreover, their antioxidant capacity and favorable effects on metabolic health, including cholesterol reduction, indicate their utility in promoting overall well-being. With their calming effects on mental health and potential to improve sleep quality, both oils represent a versatile addition to both medicinal and holistic health practices.

SUGGESTED THERAPEUTIC USE:

Inflammatory Conditions: Inflammation, pulmonary fibrosis

Infections: Bacterial infections, mycotoxin-related food safety issues (aflatoxin production)

Insect Control: Mosquito larvae infestations, fruit fly infestations

Organ and Tissue Protection: Oxidative stress-related conditions (cancer, IBD, diabetes, neurodegenerative conditions, cardiovascular disease, liver or kidney disease, metabolic syndrome, fibromyalgia, chronic fatigue syndrome, skin aging), cellular damage

Metabolic Disorders: High cholesterol, fatty liver

Mental Health Issues: Depression, anxiety

Sleep Disorders: Insomnia, poor sleep

Cancer: Cancer cell growth (breast cancer, other cancers)

Skin Issues: Impaired skin barrier function, dry skin, hyperpigmentation

Digestive Issues: poor digestion, nutrient malabsorption, liver detoxification, gut microbiome imbalance

Cognition: Reduced cognitive function

MOLECULAR AND CELLULAR TARGETS, BIOLOGICAL PATHWAYS, AND MECHANISMS:

1. Anti-inflammatory Effects

Molecular or Cellular Target/Pathway: 5-LOX.

Mechanism: Tangerine essential oil strongly inhibits the 5-LOX enzyme, reducing the synthesis of proinflammatory leukotrienes and mitigating inflammation.

2. Antimicrobial Activity

Molecular or Cellular Target/Pathway: Pathogen cell wall/membranes, aflatoxin production.

Mechanism: Tangerine and mandarin oil disrupt bacterial cell membranes, leading to increased permeability and cell lysis, effectively inhibiting pathogenic growth. Completely prevents aflatoxin B(1) production, which is crucial for controlling mycotoxin-related food safety issues.

3. Antiparasitic and Insecticidal Properties

Molecular or Cellular Target/Pathway: Mosquito larvae (e.g., *Ae. albopictus, C. pipiens*), fruit flies (*Drosophila suzukii*).

Mechanism: Tangerine and mandarin oils demonstrate potent activity against mosquito larvae, serving as natural insecticides to control mosquito populations and their spread of disease. Tangerine oil application deters fruit fly infestation on small fruits through natural repellency.

4. Antioxidant Activity

Molecular or Cellular Target/Pathway: ROS, antioxidant activity.

Mechanism: Tangerine exhibits strong antioxidant activity as evidenced in DPPH and ABTS assays, reducing oxidative stress and protecting cellular integrity.

5. Metabolic Health Benefits

Molecular or Cellular Target/Pathway: Genes regulating cholesterol metabolism (e.g., SREBP-1c, ACC, FAS).

Mechanism: Dietary supplementation with tangerine peel subcritical extract (materially similar to mandarin essential oil) downregulates cholesterol-producing genes and upregulates genes associated with bile acid synthesis, leading to reduced total cholesterol and fatty liver. SREBP-1c, or sterol regulatory element-binding protein 1c, is a protein that helps control how our body makes and uses fats, especially cholesterol. Think of SREBP-1c as a manager or director in a factory that produces cholesterol. When our body needs more cholesterol, such as when we eat fatty foods, SREBP-1c gets activated and sends signals to the cell's machinery to ramp up cholesterol production. It also helps with the creation of fats that our body uses for energy and other important functions. However, if SREBP-1c is overactive, it can lead to too much cholesterol being made, which can contribute to health issues like heart disease. So, SREBP-1c plays a crucial role in balancing cholesterol levels in our body.

6. Neuroprotective

Molecular or Cellular Target/Pathway: Brainwave activity, BDNF.

Mechanism: Inhalation of tangerine oil alters brainwave patterns, reducing alpha wave activity and promoting a theta wave state conducive to relaxation and improved sleep. It also balances elevated brain-derived neurotrophic factor (BDNF) levels. BDNF helps your brain cells grow, stay strong, and communicate better. Sleep, especially deep sleep, boosts BDNF levels, which improves memory, learning, and mood. On the other hand, not getting enough sleep lowers BDNF, making it harder to focus, remember things, and regulate emotions. Poor sleep over time can even increase the risk of brain fog, depression, and neurodegenerative diseases. So, getting good-quality sleep helps keep your brain sharp and healthy by maintaining high BDNF levels.

7. Mental Health Effects

Molecular or Cellular Target/Pathway: HPA axis, BDNF, serotonin, 5HT-1A, glucocorticoid receptors

Mechanism: Inhalation of mandarin essential oil in reserpine-treated mice balances HPA axis function, enhancing neurotransmitter production and promoting mental well-being. The oil restores the healthy expression level of 5-HT1A receptors in the brain after disruption by reserpine—a drug that has been used to induce a depression-like state in preclinical research. The healthy expression and function of 5-HT1A receptors is crucial in the treatment of mental illness, and they are a potential target for depression. The oil also increases the expression of glucocorticoid receptors, which are part of the hypothalamic-pituitary-adrenal (HPA) axis. The HPA axis is a key system involved in the body's response to stress, and its dysfunction is commonly observed in depression. Glucocorticoid receptors (GR) are like the brakes on your stress response. When you're stressed, your body releases hormones like cortisol. GR helps to regulate this response by telling your body to calm down when the stress has passed. In depression, this system can become overactive, leading to too much stress hormone. By increasing the activity of GR, treatments like mandarin oil can help restore the balance, preventing the stress response from going into overdrive and contributing to a better mood. Lastly, mandarin oil elevates the expression of BDNF. BDNF is essential for neurogenesis and synaptic plasticity, playing a vital role in the survival of dopaminergic and serotonergic neurons.

8. Cancer Prevention and Treatment

Molecular or Cellular Target/Pathway: Cytotoxicity to tumor cells.

Mechanism: Active compounds in tangerine and mandarin oil demonstrate significant cytotoxic effects against tumor cells, supporting its role as an anticancer agent.

9. Skin Health Benefits

Molecular or Cellular Target/Pathway: Tyrosinase, keratinocytes, skin barrier function, and skin microbiome.

Mechanism: The oil prevents hyperpigmentation by inhibiting tyrosine intracellularly and extracellularly, thus showing potential for skin care applications. Intracellular inhibitors work deep within the cell, while extracellular inhibitors work on the cell's surface. When tyrosinase is inhibited, the conversion of tyrosine to melanin is reduced or stopped. This results in less melanin production and lighter skin or improved skin tone. It also enhances skin barrier integrity and hydrates the skin when combined with tea tree, lavender, and eucalyptus, demonstrating its utility in skincare formulations. Tangerine deeply penetrated the skin—better than the other oils tested.

10. Bile Acid Synthesis (Digestion, Nutrient Absorption, Liver Detoxification, and Gut Microbiome Regulation)

Molecular or Cellular Target/Pathway: Genes related to bile acid production (LXRα, CYP7A1, and CYP27A1).

Mechanism: The oil upregulates expression of LXRα, CYP7A1, and CYP27A1, which helps our body produce more bile acids, leading to multiple health benefits. Bile acids are like special detergents that our body uses to break down fats in the food we eat, which improves digestion and helps absorb important nutrients. Specifically, increased bile acid production improves fat digestion and absorption of fat-soluble vitamins (A, D, E, K). This is important for issues like fatty stool (steatorrhea) and malabsorption syndromes. Additionally, these bile acids play a key role in liver detoxification by helping to remove toxins and waste products from the body. They also support a healthy balance of gut bacteria by sending signaling molecules that regulate the gut microbiome, which is essential for good digestive health. Lastly, bile acids influence neurotransmitter balance and neuroprotection, which may also contribute to cognitive function and mental clarity. So, by boosting these genes, our body gets better at breaking down food, cleaning out toxins, maintaining a balanced gut microbiome, and influencing neurotransmitters, all of which are important for overall health.

11. Prevention of Pulmonary Fibrosis

Molecular or Cellular Target/Pathway: Oxidative stress and connective tissue growth factor (CTGF) protein signaling.

Mechanism: Prevents collagen deposition in lung tissue and downregulates CTGF gene expression, which plays a role in fibrosis development. CTGF is a protein that plays a significant role in the process of pulmonary fibrosis, a condition where the lungs become scarred and stiff. When the lungs are injured or stressed, CTGF gets activated and signals other cells to produce more collagen and other substances that contribute to scar tissue formation. While this can be a helpful response to repair damage, in pulmonary fibrosis, this signaling goes into overdrive. As a result, excessive scarring occurs in the lung tissue, making it harder for the lungs to function properly, leading to breathing difficulties and reduced oxygen flow to the body.

TEA TREE

Tea tree (*Melaleuca alternifolia*) essential oil, derived from the leaves of the tree, is renowned for its broad-spectrum antimicrobial properties and therapeutic effects. Numerous studies highlight its effectiveness in preventing dental plaque, inhibiting bacterial and fungal infections, and promoting wound healing. The active compounds, particularly terpinen-4-ol, are recognized for their ability to disrupt the cellular functions of pathogens, leading to their elimination. Additionally, tea tree oil has been shown to alleviate inflammation and possess potential applications in treating skin conditions such as acne and psoriasis. By modulating immune responses and promoting cellular repair, tea tree essential oil serves as a versatile and natural remedy for various health issues, making it an integral component of many therapeutic regimes.

SUGGESTED THERAPEUTIC USE:

Infections: Oral infections, bacterial infections, plaque formation

Fungal Infections: Fungal infections, onychomycosis (nail infections)

Inflammatory Conditions: Acute inflammation, psoriasis

Wounds: Wounds, infected wounds

Skin Conditions: Acne, allergic reactions (histamine-mediated), erythema

Immune Dysfunction: Weakened immune function

Viral Infections: Influenza A/PR/8 H1N1, HSV-1, HSV-2, other enveloped viruses

Insect Control: Lice, ticks, insect infestations

Oral Hygiene: Plaque, gingivitis

MOLECULAR AND CELLULAR TARGETS, BIOLOGICAL PATHWAYS, AND MECHANISMS:

1. Antimicrobial Activity

Molecular or Cellular Target/Pathway: Pathogen cellular integrity.

Mechanism: Tea tree essential oil effectively reduces the levels of oral and other bacteria, limiting plaque formation on teeth and supporting overall oral health by disrupting cellular integrity. The oil's compounds permeate bacterial cell membranes, leading to cell lysis, leakage of cytoplasmic contents, and inhibition of metabolic processes necessary for survival.

2. Antifungal Effects

Molecular or Cellular Target/Pathway: Pathogen cell integrity and hyphae, spore germination, ergosterol synthesis, biofilm formation.

Mechanism: Tea tree oil demonstrates strong antifungal action by altering fungal cell membrane permeability and preventing spore germination, effectively disrupting growth and biofilm formation. The oil also interferes with ergosterol synthesis and damages hyphae (thread-like structures that make up the fungal network), compromising the structural integrity of fungal cells, making them vulnerable and unable to survive. It is particularly effective against dermatophytes that infect the skin, hair, and nails. Low concentrations of tea tree oil inhibit fungal growth, providing a potential alternative treatment for nail infections.

3. Anti-inflammatory Effects

Molecular or Cellular Target/Pathway: Proinflammatory cytokines (e.g., IL-1β, IL-6), MAPK signaling pathways.

Mechanism: Tea tree oil inhibits the production of proinflammatory molecules through the modulation of MAPK signaling pathways, reducing the inflammatory response in conditions such as psoriasis. The MAPK signaling pathway plays a significant role in regulating the release of proinflammatory cytokines. This pathway is activated in response to various extracellular signals such as stress, growth factors, and inflammatory stimuli. The MAPK pathway consists of several key components, including ERK (extracellular signal-regulated kinase), JNK (c-Jun N-terminal kinase), and p38 MAPK. In response to proinflammatory stimuli, these MAPKs are phosphorylated—a process in which a phosphate group (a small chemical structure made up of phosphorus and oxygen) is added to a molecule, usually a protein, to activate or deactivate that molecule—and activated. Once activated, MAPK pathways can translocate to the nucleus and activate various transcription factors (like NF-κB, AP-1, and others) that are responsible for the transcription of genes encoding proinflammatory cytokines (such as TNF-α, IL-6, IL-1β, and others). This process increases the synthesis and release of these cytokines from immune cells. The increase in cytokine levels can have several physiological effects, including promoting inflammation, attracting immune cells to the site of infection or injury, and facilitating the overall immune response. However, excessive or dysregulated cytokine production can lead to chronic inflammation and is implicated in various inflammatory diseases.

4. Wound-Healing Properties

Molecular or Cellular Target/Pathway: Various growth factors and immune cells, wound pathogens.

Mechanism: Tea tree oil promotes wound healing by increasing collagen synthesis and attracting immune cells to the site of injury, thus accelerating tissue regeneration and restoring skin integrity. Topical application of tea tree oil effectively reduces bacterial load in infected wounds, promoting faster healing rates compared to standard treatments.

5. Acne Treatment and Skin Health

Molecular or Cellular Target/Pathway: Bacteria associated with acne (*C. acnes*) and inflammatory mediators.

Mechanism: Tea tree oil displays antimicrobial activity against acne-associated bacteria while also reducing inflammation, resulting in fewer acne lesions and less erythema. Acne lesions are characterized by inflammation. Tea tree oil can help reduce this inflammation, which contributes to the redness and swelling associated with pimples. Tea tree may also regulate sebum production.

Molecular or Cellular Target/Pathway: H1 histamine receptors, substance P-induced microvascular changes, and NO signaling.

Mechanism: Tea tree oil is a promising solution for allergic skin reactions as well. Tea tree oil demonstrably soothes inflamed skin by mitigating histamine-mediated allergic skin eruptions. This effect is largely attributed to terpinen-4-ol, its primary active constituent. Terpinen-4-ol works through several mechanisms: it interacts with H1 histamine receptors on endothelial cells, thereby regulating vasodilation and plasma leakage; it blocks substance P-induced microvascular changes, a key component of allergic inflammation; and it modulates sensory C fibers, reducing the release of inflammatory neuropeptides. Additionally, terpinen-4-ol inhibits nitric oxide (NO) release, which further prevents excessive vasodilation and reduces histamine-induced redness and swelling. Inhibits histamine-induced microvascular permeability, reducing swelling and redness (wheal and flare). Modulates nitric oxide (NO) signaling, which plays a role in vascular relaxation and inflammation. Reduces substance P-mediated vasodilation and inflammation, dampening allergic skin reactions. Suppresses neurogenic inflammation caused by nerve activation during allergic responses.

6. Immune Modulation

Molecular or Cellular Target/Pathway: Immune cells, particularly leukocytes.

Mechanism: Tea tree oil activates immune responses. It enhances the activity of polymorphonuclear leukocytes, promoting pathogen clearance and supporting immune function during infections. Polymorphonuclear leukocytes (PMNs), also known as neutrophils, are a type of white blood cell that play a vital role in the immune system. They are the body's first line of defense against infections, particularly bacterial and fungal infections. When an infection occurs or when tissues are damaged, PMNs quickly move to the site of injury or infection, where they help fight off pathogens. They do this by engulfing (consuming) the invaders through a process called phagocytosis, and they release enzymes and other substances that help kill and digest the pathogens. Additionally, PMNs can release signaling molecules that attract other immune cells to the area, amplifying the immune response.

7. Viral Inhibition

Molecular or Cellular Target/Pathway: Influenza virus is influenza A/PR/8 H1N1 and other enveloped viruses (HSV-1, HSV-2).

Mechanism: Tea tree oil interferes with viral envelope integrity, preventing bacterial adhesion and replication without harming host cells. It may also interfere with viral structures that are important for the early steps of viral infection, which could include adsorption (attachment of the virus to the host cell) and entry.

8. Insecticidal Properties

Molecular or Cellular Target/Pathway: Toxicity to various insect species such as lice and ticks.

Mechanism: Tea tree oil can both repel and kill insects. The essential oil exerts contact toxicity and repellent effects, disrupting physiological functions and providing a natural alternative to synthetic insecticides.

9. Dental Hygiene

Molecular or Cellular Target/Pathway: Bacterial biofilms and proinflammatory mediators.

Mechanism: Tea tree oil has a long history of use to promote oral health. Tea tree oil-based mouthwashes significantly reduce plaque and gingivitis, demonstrating potential as an adjunctive treatment in dental hygiene. Tea tree oil, particularly its component terpinen-4-ol, has been shown to inhibit the growth of various oral bacteria, including those associated with plaque formation and gingivitis. Gingivitis is characterized by inflammation of the gums. Tea tree oil suppresses inflammatory responses to reduce the redness, swelling, and bleeding associated with gingivitis.

THYME

Thyme essential oil, derived from the *Thymus vulgaris* or *Th. zygis* plant, is noted for its potent antimicrobial and anticancer properties, as well as its ability to modulate inflammation and support metabolic health. Its composition can vary greatly with thymol, carvacrol, linalool, thujanol (thuyanol-4)—considered a gentler chemotype, and even geraniol or alpha-terpineol being dominant. The most common CT available commercially is the thymol CT. Enriched with active compounds such as thymol and para-cymene, thyme oil demonstrates significant efficacy against various pathogenic microorganisms, including drug-resistant bacteria and fungi. Furthermore, research shows that thyme essential oil induces apoptotic processes in cancer cells and offers protective effects against oxidative stress and inflammation. Its diverse mechanisms of action suggest thyme oil as

a valuable asset in both therapeutic and preventive healthcare strategies.

SUGGESTED THERAPEUTIC USE:

Infections: Bacterial infections, drug-resistant bacterial infections, biofilm-related infections (chronic wound infections, catheter-associated UTIs, heart valve infections, chronic sinusitis, cystic fibrosis, dental plaque, periodontal disease, osteomyelitis, conjunctivitis, implant-associated infections, ventilator-associated pneumonia, dialysis-associated infections), fungal infections, fungal biofilms (candidiasis, aspergillosis, cryptococcosis, onychomycosis, fungal conjunctivitis, endocarditis, medical equipment—catheters, implants—infections)

Cancer: Liver cancer, breast cancer, glioblastoma, cancer cell proliferation

Inflammatory Conditions: Chronic inflammatory conditions (rheumatoid arthritis, IBD, psoriasis, COPD, asthma, lupus, eczema, gout, chronic sinusitis, fibromyalgia, chronic fatigue syndrome, periodontitis), conditions with elevated MPO (rheumatoid arthritis, IBD, lupus, cardiovascular diseases, infections, diabetes, obesity, cancers, asthma, COPD), elevated proinflammatory cytokines (autoimmune conditions, lupus, multiple sclerosis, infections, cancers, metabolic disorders, neurological disorders, sepsis)

Organ and Tissue Protection: Oxidative stress-related conditions (cancer, IBD, diabetes, neurodegenerative conditions, cardiovascular disease, liver or kidney disease, metabolic syndrome, fibromyalgia, chronic fatigue syndrome, skin aging), cellular damage

[189]

Cardiovascular Issues: High LDL cholesterol, cardiovascular disease risk

Gastrointestinal Issues: Gut infections, gastrointestinal disorders

Neuroprotection: Neurodegenerative disorders, cognitive decline, memory problems

Insect Control: Pest infestations

Viral Infections: Viral infections (HSV-1, HSV-2)

MOLECULAR AND CELLULAR TARGETS, BIOLOGICAL PATHWAYS, AND MECHANISMS:

1. Antimicrobial Activity

Molecular or Cellular Target/Pathway: Pathogen cell walls/membranes, biofilm formation.

Mechanism: Thyme essential oil disrupts bacterial cell membranes, resulting in cell lysis and death. It also inhibits the biofilm formation process, which is crucial for the persistence of infections. Thyme essential oil demonstrates strong bactericidal and antibacterial activity that can effectively combat drug-resistant pathogens through direct cell membrane disruption and inhibition of pathogen metabolic pathways.

2. Anticancer Properties

Molecular or Cellular Target/Pathway: Proapoptotic cancer cell pathways (mitochondrial, caspase, EGFR, JAK-STAT).

Mechanism: Carvacrol and thymol, both key phenols in thyme oil, trigger apoptosis via the mitochondrial pathway, activating caspases that lead to programmed cell death. This reduces tumor growth and frequency in animal studies. Thyme oil regulates critical cell signaling pathways involved in cancer progression (e.g., EGFR, JAK-STAT), resulting in reduced cell viability and tumor growth.

3. Anti-inflammatory Properties

Molecular or Cellular Target/Pathway: Myeloperoxidase (MPO), proinflammatory cytokines (e.g., IL-6, TNF-α).

Mechanism: Thyme essential oil inhibits the activity of proinflammatory enzymes and reduces the secretion of inflammatory mediators, alleviating conditions characterized by chronic inflammation. MPO is an enzyme released by white blood cells, particularly neutrophils, during inflammation. Essentially, MPO helps these cells create powerful oxidants to fight off invaders like bacteria. However, when there's excessive or prolonged inflammation, MPO can contribute to tissue damage by these same oxidants. So, while it's a tool for defense, too much MPO activity can exacerbate inflammatory conditions by causing collateral damage to healthy cells and tissues.

4. Antioxidant Activity

Molecular or Cellular Target/Pathway: Reactive oxygen species (ROS), LDL cholesterol oxidation.

Mechanism: Thyme essential oil effectively scavenges free radicals and prevents the oxidation of LDL cholesterol, potentially reducing the risk of cardiovascular diseases.

5. Antifungal Effects

Molecular or Cellular Target/Pathway: Pathogen cell walls/membranes, biofilm formation.

Mechanism: Thyme oil demonstrates fungicidal activity by altering cell membrane permeability and inhibiting the growth of filamentous forms, thus reducing fungal colonization and biofilm formation.

6. Gastrointestinal Health

Molecular or Cellular Target/Pathway: Gut pathogens (*H. pylori, E. coli*).

Mechanism: Thyme essential oil may inhibit pathogenic bacteria in the gastrointestinal tract, promoting gut health and potentially alleviating symptoms of gastrointestinal disorders.

7. Neuroprotective Effects

Molecular or Cellular Target/Pathway: Acetylcholinesterase (AChE).

Mechanism: Thyme's constituents, like thymol and carvacrol, may protect against neurodegenerative diseases. Both carvacrol and thymol inhibit AChE activity, preventing the breakdown of acetylcholine, which is crucial for cognitive function. This suggests potential benefits for memory and neuroprotection.

8. Insecticidal Activity

Molecular or Cellular Target/Pathway: Insect cellular respiration and development.

Mechanism: Thyme essential oil exhibits insecticidal effects that disrupt various physiological processes in pests, thereby aiding in pest control in agricultural settings.

9. Antiviral Activity

Molecular or Cellular Target/Pathway: Various viral proteins (e.g., HSV-1, HSV-2).

Mechanism: Thyme essential oil demonstrates potent antiviral properties. It shows virucidal activity by preventing virus attachment and replication within healthy cells, showcasing its potential as a natural antiviral agent. Thyme oil exhibits virucidal activity, meaning it can directly inactivate the virus. It may also interact with the viral envelope, potentially disrupting the viral envelope or masking viral components necessary for replication and infectivity.

10. Cardioprotective

Molecular or Cellular Target/Pathway: Calcium channels and handling, proinflammatory cytokines (TNF-α, IL-6, IL-1β), oxidative stress damage, mitochondrial function, caspase-8, caspase-9, FAS, Bcl-xL genes expression.

Mechanism: Thyme oil, largely due to its active ingredient thymol, offers several heart-protecting benefits that can positively impact cardiovascular health. One of the key ways thymol helps is by promoting vasorelaxation, which means it helps blood vessels relax and widen, making it easier for blood to flow. This action likely occurs through a process that regulates calcium levels within the heart muscle, effectively reducing tension and enhancing relaxation.

Additionally, thymol assists in managing how calcium is handled in heart cells, leading to improved heart function. Moreover, thymol possesses powerful anti-inflammatory properties that can reduce inflammation, which is essential because chronic inflammation can lead to various heart diseases. Thymol has been shown to inhibit the release of certain harmful enzymes (lysosomal enzymes) and decrease the levels of proinflammatory substances (TNF-α, IL-6, IL-1β) in heart tissue. Lysosomal enzymes play a crucial role in heart function by maintaining cellular homeostasis, breaking down macromolecules, and preventing the accumulation of toxic materials. However, in certain circumstances, like during a heart attack, oxygen deprivation (ischemia) and subsequent reoxygenation (reperfusion) generate oxidative stress, which can damage lysosomal membranes, causing them to burst. When this happens, the cleaning enzymes spill out into the cell, and instead of cleaning up, these enzymes start breaking down important parts of the heart cell, causing it to die. Inhibiting lysosomal enzymes during a heart attack can protect cardiac cells from excessive degradation, inflammation, and oxidative stress, potentially improving heart recovery and function after injury. However, timing and degree of inhibition are critical, as lysosomes are also essential for cellular repair and long-term recovery. Furthermore, thymol also plays a protective role against damage from chemotherapy drugs, such as doxorubicin, by combating oxidative stress and inflammation that can harm heart cells. It normalizes the activities/levels of key antioxidants in the heart (SOD, CAT, GPx), important enzymes (isocitrate dehydrogenase, malate dehydrogenase, α-ketoglutarate dehydrogenase, and NADH-dehydrogenases, cytochrome c oxidase), and energy molecules (ATP). Another significant benefit of thymol is its ability to support mitochondrial function, which is vital for energy production in heart cells. By reducing oxidative stress and normalizing important biochemical activities within mitochondria, thymol helps maintain the health of heart cells. Lastly, thymol also protects against cell death caused by oxidative stress, thereby preserving heart cell integrity and function. Overall, the diverse effects of thymol from thyme oil highlight its potential as a natural agent for promoting heart health and protecting against cardiovascular issues.

11. Pain Relief and Antispasmodic Effects

Molecular or Cellular Target/Pathway: Voltage-gated sodium channels, calcium handling in skeletal muscle, myosin ATPase activity and muscle contraction, α1-adrenergic, α2-adrenergic, and β-adrenergic receptors, GABA receptors.

Mechanism: Thymol, a compound found in thyme oil, has several effects that help relieve pain and reduce muscle spasms. One of its primary actions is blocking voltage-gated sodium channels, which are important for sending signals in nerve and muscle cells. Thymol binds more effectively to these channels when they are in an inactive state, similar to how the local anesthetic lidocaine works, helping to numb pain. In skeletal muscles, thymol improves the release of calcium from storage areas inside muscle cells when the cells are stimulated. This means that when a muscle is triggered to contract, thymol helps it to do so more efficiently, enhancing muscle function. At higher concentrations, thymol can keep certain calcium channels open longer, which could further aid in healthy muscle contraction. Thymol also influences myosin, a protein involved in muscle contractions. It activates myosin's ability to use energy (ATPase activity), which is crucial for muscle movement. However, at the same time, it can reduce the overall strength of muscle contractions, indicating a complex effect on how muscles work together. In smooth muscles, which control many involuntary movements in the body, thymol can alter contraction patterns. It acts on specific receptors (α1-adrenergic, α2-adrenergic, and β-adrenergic receptors) that help manage muscle contractions, providing a way to ease excessive tightening or spasms. Additionally, thymol interacts with GABA receptors in the nervous system, enhancing their activity. This interaction might contribute to its pain-relieving effects, as these receptors play a role in calming nerve activity. Thymol also changes the structure of lipid membranes, which could be part of how it produces its anesthetic-like properties. In summary, thymol delivers pain relief and helps reduce muscle cramps through various actions, including blocking nerve signals, enhancing calcium release in muscles, affecting muscle contraction dynamics, and calming nerve activity.

TURMERIC

Turmeric essential oil, derived from the *Curcuma longa* plant (most commonly the roots, but also the leaves), presents a broad spectrum of therapeutic potential, extending beyond its culinary uses. Research reveals its efficacy in combating parasitic infections like toxoplasmosis, inhibiting viral proteases associated with dengue fever, and demonstrating potent antibacterial activity against multidrug-resistant strains. Furthermore, turmeric oil exhibits insecticidal and acaricidal properties, effectively repelling and eliminating pests like fire ants and cattle ticks. Notably, its ability to rescue fibroblasts from cellular senescence and reduce oxidative stress suggests significant anti-aging potential. Studies also highlight its roles in wound healing, blood sugar regulation, and cholesterol management. The oil's anti-inflammatory and neuroprotective effects indicate promise in addressing conditions ranging from arthritis to neurodegenerative disorders.

SUGGESTED THERAPEUTIC USE:

Cancer: Colon, breast, liver, prostate, cervical, oral submucous fibrosis (pre-cancerous condition)

Infections: Parasitic (toxoplasmosis, trypanosomiasis, hydatid cysts), viral (dengue), bacterial, fungal

Immune System Support: General immune system modulation

Inflammatory Conditions: Chronic inflammatory conditions (rheumatoid arthritis, IBD, psoriasis, COPD, asthma, lupus, eczema, gout, chronic sinusitis, fibromyalgia, chronic fatigue syndrome, periodontitis)

Pain Management: Joint pain, general pain relief

Neurological Disorders: Stroke recovery, neurodegenerative diseases (Alzheimer's disease, Parkinson's disease, Huntington's disease), multiple sclerosis, seizures

Mental Health Issues: Anxiety, depression

Metabolic Disorders: Diabetes, high cholesterol

Skin Conditions: Eczema, psoriasis, wounds, skin aging, skin infections

Gastrointestinal Disorders: IBD, ulcerative colitis, GI ulcers

Oral Health: Periodontitis, dental plaque, dental cavities

Male Reproductive Health: Benign prostatic hyperplasia

Pest Problems: Insect repellency, tick infestations

MOLECULAR AND CELLULAR TARGETS, BIOLOGICAL PATHWAYS, AND MECHANISMS:

1. Antiparasitic, Antiviral, and Antimicrobial Effects

Molecular or Cellular Target/Pathway: *T. gondii* tachyzoites, dengue virus protease (NSB-NS3), biofilm formation, bacterial cell walls/membranes, fungal cell walls/membranes.

Mechanisms: Turmeric essential oil directly inhibits *T. gondii* parasite replication, dengue viral protease, and disrupts microbial cell membrane integrity, and inhibition of biofilm formation.

2. Anti-inflammatory and Analgesic Effects

Molecular or Cellular Target/Pathway: 5-LOX enzyme, inflammatory cytokines (TNF-α, IL-6, IL-1β), NF-κB, COX-2, p38 MAPK.

Mechanisms: Turmeric essential oil's inhibition of inflammatory enzyme activity (5-LOX, COX-2), reduction of proinflammatory cytokine production (TNF-α, IL-6, IL-1β), and modulation of inflammatory signaling pathways (NF-κB, p38 MAPK) leads to a healthy inflammatory response.

3. Antioxidant and Anti-aging Effects

Molecular or Cellular Target/Pathway: ROS, fibroblasts, cellular senescence, AhR.

Mechanisms: Turmeric oil scavenges free radicals, protecting cells from oxidative stress. It also promotes fibroblast survival and function, which are the primary cells responsible for producing collagen and elastin, the structural proteins that give skin its firmness, elasticity, and youthful appearance. As we age, fibroblast activity declines, leading to reduced collagen and elastin production, resulting in wrinkles, sagging skin, and loss of skin tone. By supporting fibroblast function, turmeric oil can help maintain or even increase the production of these vital proteins, restoring skin's structural integrity. Fibroblasts also maintain the extracellular matrix (ECM), the network of proteins and other molecules that provides structural support to tissues. A healthy ECM is essential for skin hydration, wound healing, and overall tissue health. By promoting fibroblast function, turmeric oil helps maintain a healthy ECM, preventing tissue degeneration and promoting a youthful appearance. Additionally, turmeric oil reduces cellular senescence—a process where cells stop dividing and contribute to tissue aging. Turmeric oil has been shown to rescue fibroblasts from cellular senescence, meaning it can help keep these cells functioning for a longer period of time, slowing down the aging process. Lastly, the oil acts on the transcription factor aryl hydrocarbon receptor (AhR), which plays a crucial role in regulating gene expression in response to environmental chemicals. AhR can interact with and influence the expression of genes that are involved in regulating the body's internal clock (circadian rhythm). This means that AhR can affect the timing of various physiological processes that follow a 24-hour cycle, such as sleep-wake cycles, hormone release, and metabolism. Turmeric inhibits AhR, which can be beneficial in reducing the harmful effects of certain environmental toxins.

4. **Neuroprotective Effects**

Molecular or Cellular Target/Pathway: Neural stem cells, microglia, neuroinflammation, brain glucose metabolism.

Mechanisms: The primary constituent in turmeric oil, ar-Turmerone stimulates neural stem cell production, inhibits microglia activation, reduces neuroinflammation, and normalizes brain glucose metabolism. It helps the brain produce more neural stem cells, which are important for repairing and regenerating brain tissue. By reducing the activation of microglia (the brain's immune cells that can cause inflammation if overactive), ar-Turmerone helps lower inflammation in the brain, which is linked to many neurological diseases, including depression. Additionally, it helps keep brain glucose metabolism balanced, ensuring the brain gets the energy it needs to function properly. Together, these actions could support better brain health and reduce the risk or severity of neurological diseases, helping us maintain cognitive function as we age.

5. **Anticancer Effects**

Molecular or Cellular Target/Pathway: Cancer cell proliferation, apoptosis pathways, angiogenesis.

Mechanisms: By inducing apoptosis in cancer cells, inhibiting tumor growth, and suppressing angiogenesis, turmeric attacks multiple aspects of cancer development, progression, and metastasis.

6. **Metabolic Regulation**

Molecular or Cellular Target/Pathway: Alpha-glucosidase, alpha-amylase, PPAR-α, LXRa, lipid metabolism genes.

Mechanisms: Turmeric oil inhibits carbohydrate-digesting enzymes (alpha-glucosidase, alpha-amylase), modulates lipid metabolism genes, and reduces cholesterol synthesis. PPARs and liver X receptors (LXRs) are nuclear receptor proteins that play critical roles in the regulation of metabolism, including glucose and lipid metabolism. By activating PPARs, particularly PPAR-gamma and PPAR-alpha, there is an enhancement of glucose uptake and metabolism in peripheral tissues, along with improved lipid profiles due to increased fatty acid oxidation and decreased circulating triglycerides. LXR activation increases the expression of genes involved in cholesterol efflux, such as ATP-binding cassette transporters. This helps to remove excess cholesterol from cells and transport it to the liver for excretion (bile), thereby reducing overall cholesterol levels. LXRs also regulate lipid synthesis and metabolism. By promoting pathways that enhance fatty acid synthesis and storage, LXRs can support stable energy storage and utilization, which helps to manage blood sugar levels. They can also influence insulin sensitivity positively.

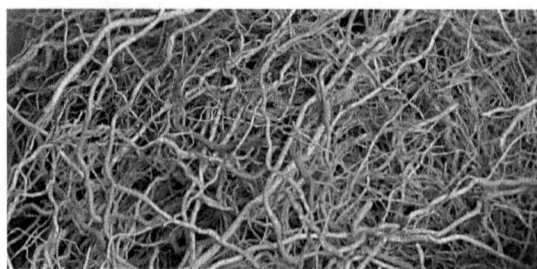

VETIVER

Extracted from the roots of the plant, vetiver (*Vetiveria zizanioides*) essential oil is renowned for its diverse therapeutic properties, including potent antioxidant, antimicrobial, and anti-inflammatory effects. Rich in bioactive sesquiterpene compounds such as khusimol and (E)-Isovalencenol, and vetivones, vetiver oil exhibits significant efficacy in preventing oxidative damage, modulating stress responses, and inhibiting pathogenic microorganisms. Research suggests that vetiver has applications in cancer therapy, demonstrating promise in inhibiting the growth of certain cancer cells and promoting apoptosis. Additionally, it is noted for enhancing cognitive function, improving mood, and offering insect-repellent properties. This multifaceted essential oil serves as a valuable resource in both holistic and clinical health practices.

SUGGESTED THERAPEUTIC USE:

Organ and Tissue Protection: Oxidative stress-related conditions (cancer, IBD, diabetes, neurodegenerative conditions, cardiovascular disease, liver or kidney disease, metabolic syndrome, fibromyalgia, chronic fatigue syndrome, skin aging), cellular damage, lipid peroxidation

Neuroprotection: Cognitive decline, poor attention, slow reaction time, mental fog

Mental Health Issues: Anxiety, stress, stress responses

Infections: Bacterial infections, fungal infections

Cancer: Colon cancer, breast cancer

Insect Repellency: Mosquito-borne illnesses (e.g., prevention of bites from *Ae. aegypti* and *An. minimus*)

Sleep Disorders: Disrupted sleep, excessive slow wave sleep, lack of alertness after sleep

Parasitic Infections: Protozoan infections (e.g., *T. cruzi*, *L. amazonensis*)

MOLECULAR AND CELLULAR TARGETS, BIOLOGICAL PATHWAYS, AND MECHANISMS:

1. Antioxidant Activity

Molecular or Cellular Target/Pathway: Lipid peroxidation, free radicals, endogenous antioxidants (SOD, GPx, CAT).

Mechanism: Vetiver essential oil acts as a free radical scavenger, reducing oxidative stress and preventing damage to cellular lipids, which is crucial for maintaining cellular integrity and function. Vetiver exhibits potent lipid peroxidation inhibitory activity, maintaining cellular glutathione (GSH) levels. Decreased malondial-dehyde—a cellular oxidant—levels were observed, as well as restoration of SOD, GPx, and CAT.

2. Neuroprotective Effects

Molecular or Cellular Target/Pathway: Central nervous system pathways—olfactory pathways in the brain, including the cerebral cortex, hypothalamus, and amygdala.

Mechanism: Inhalation of vetiver oil modulates brainwave activity, improving attention and reaction time, indicating its potential benefits in cognitive enhancement and mental clarity. Vetiver increased average LF/HF ratio, which was higher under low-dose conditions. The LF/HF ratio refers to the ratio of low-frequency to high-frequency components and a measure of balance between the sympathetic and parasympathetic nervous systems. These components reflect the balance between the two branches of the autonomic nervous system: the sympathetic (LF) and parasympathetic (HF) nervous systems. A high LF/HF ratio indicates that the sympathetic nervous system is more active, while a low LF/HF ratio indicates that the parasympathetic nervous system is more active. In other words, the LF/HF ratio is a measure of sympathovagal balance. A higher LF/HF ratio suggests a higher level of stress, anxiety, or physical exertion, as the sympathetic nervous system takes precedence. Conversely, a lower LF/HF ratio indicates a more relaxed state, where the parasympathetic nervous system is more dominant.

3. Anxiety and Stress Resilience Properties

Molecular or Cellular Target/Pathway: Central amygdaloid nucleus.

Mechanism: Vetiver can help relieve anxiety and improve stress resilience. Vetiver essential oil alters neuronal activity in the amygdala, a region associated with fear and anxiety, thus exhibiting anxiolytic effects that can help alleviate stress.

4. Antimicrobial Activity

Molecular or Cellular Target/Pathway: Pathogen cell walls/membranes and metabolism.

Mechanism: Vetiver oil demonstrates bactericidal activity by disrupting cell membrane integrity and inhibiting the metabolic processes of various pathogenic organisms, effectively preventing infection and proliferation.

5. Anticancer Properties

Molecular or Cellular Target/Pathway: Cancer cell receptors and pathways (e.g., CB2 receptors).

Mechanism: Vetiver essential oil with 15.4% beta-caryophyllene promotes apoptosis in colon and breast cancer cells through specific receptor binding and molecular interactions, leading to decreased tumor growth and enhanced cancer cell death. CB2 receptors are expressed not only in the immune system but also in some types of cancer cells. Activation of CB2 can influence various processes in cancer, such as cell proliferation, apoptosis (programmed cell death), invasion, and metastasis. Some studies indicate that CB2 activation might inhibit tumor growth and reduce inflammation in the tumor microenvironment, potentially providing therapeutic benefits.

6. Insect Repellent Activity

Molecular or Cellular Target/Pathway: Olfactory receptors in insects.

Mechanism: Vetiver essential oil exhibits repellent and lethal effects on mosquito vectors, particularly *Ae. aegypti* and *An. minimus*, by interfering with their sensory perceptions and potentially causing physiological distress.

7. Effects on Sleep Patterns

Molecular or Cellular Target/Pathway: Brainwave activity (decreased alpha and beta1 activity, and increased gamma activity), circadian rhythm.

Mechanism: Inhalation of vetiver oil influences EEG patterns, decreasing slower wave sleep while enhancing alertness and reaction times, thus suggesting its role in sleep regulation and mental alertness. Circadian rhythms significantly influence our sleep stages, alertness, and the quality of our sleep. When inhaled, vetiver oil appears to alter brain activity by decreasing slower-wave sleep, which is typically associated with deep, restorative sleep. Deep sleep is important for physical recovery and cognitive function, and its reduction could have implications for overall health. Additionally, the increase in alertness and quicker reaction times when using vetiver oil suggests it may have stimulant-like properties that counteract feelings of sleepiness, potentially helping to regulate the balance between rest and wakefulness. This is especially relevant during times when people might typically feel groggy according to their circadian cycle, such as during the afternoon slump. By influencing EEG patterns and altering the depth of sleep and levels of alertness, vetiver oil may interact with circadian rhythms, impacting sleep quality and daytime performance.

8. Antiparasitic Effects

Molecular or Cellular Target/Pathway: *T. cruzi, L. amazonensis* parasites survival and replication.

Mechanism: Vetiver oil demonstrates significant activity against both the promastigote and amastigote forms of these parasites, affecting their survival and replication processes.

WINTERGREEN

Wintergreen essential oil, derived from *Gaultheria* spp. plants, is well-known for its analgesic, anti-inflammatory, and antimicrobial properties, primarily attributed to its high content of methyl salicylate. Research indicates that wintergreen oil acts similarly to aspirin, providing pain relief and anti-inflammatory effects without the gastrointestinal side effects commonly associated with traditional NSAIDs. Indeed, a single drop of high-quality wintergreen essential oil contains approximately the same amount of salicylate as a baby aspirin, while four drops provide a comparable salicylate content to that found in a 325 mg aspirin tablet. Its application extends to numerous areas, including headache relief, treatment for cutaneous leishmaniasis, and as a biocide against certain pests. The mechanisms of action involve inhibition of proinflammatory mediators, antioxidant effects, and direct toxicity to pathogenic organisms, making wintergreen oil a versatile compound in both medicinal and agricultural settings.

<u>Suggested Therapeutic Use:</u>

Anti-inflammatory Activity: Acute inflammation, chronic inflammatory conditions (rheumatoid arthritis, IBD, psoriasis, COPD, asthma, lupus, eczema, gout, chronic sinusitis, fibromyalgia, chronic fatigue syndrome, periodontitis), swelling, conditions involving proinflammatory cytokines (autoimmune conditions, lupus, multiple sclerosis, infections, cancers, metabolic disorders, neurological disorders, sepsis)

Pain: Headaches, pain, musculoskeletal pain, arthritis, abdominal pain, abdominal contractions, bursitis, tendonitis, muscle strains or sprains, back pain, sports injuries

Organ and Tissue Protection: Oxidative stress-related conditions (cancer, IBD, diabetes, neurodegenerative conditions, cardiovascular disease, liver or kidney disease, metabolic syndrome, fibromyalgia, chronic fatigue syndrome, skin aging), cellular damage

Infections: Bacterial infections, fungal infections

Insect Control and Repellency: Infestations by pests such as granary weevils and spotted wing drosophila

Parasitic Infection: Cutaneous leishmaniasis

<u>Molecular and Cellular Targets, Biological pathways, and Mechanisms:</u>

1. Anti-inflammatory Activity

Molecular or Cellular Target/Pathway: 5-LOX, COX, proinflammatory cytokines (TNF-α and IL-1), pain receptors.

Mechanism: Salicylate glucoside (SG) from the wintergreen plant, not oil, inhibits the activity of 5-LOX, thereby reducing leukotriene synthesis, and limits nitric

[199]

oxide production, leading to decreased inflammation. Both methyl salicylate (MS) and SG are derived from salicylic acid and exhibit anti-inflammatory and analgesic properties. SG is a glycoside, meaning it has a sugar moiety (glucose) attached to the salicylate part. MS is an organic ester formed during distillation of the leaves from gaultherin. MS is lipophilic, which allows it to be readily absorbed through the skin. It is also actively metabolized in the body, leading to the release of salicylic acid. SG is more polar, making it less absorbable through the skin. It may require enzymatic cleavage to release the active salicylate for similar effects. SG can be precursors to methyl salicylate. Through enzymatic or chemical processes, the sugar component can be removed, and the salicylate can be converted to methyl salicylate. Essentially, the glucoside form is a storage and transport form, while the methyl salicylate is a more active form. While both are related to salicylic acid, they don't have identical properties. MS has direct therapeutic activity, while SG's therapeutic properties depend on breakdown. Methyl salicylate, when absorbed through the skin, is metabolized into salicylic acid. Salicylic acid inhibits COX enzymes, thereby reducing prostaglandins and mitigating inflammatory responses. Salicylic acid can inhibit the activation of macrophages, reducing their production of proinflammatory cytokines like TNF-α and IL-1. It can affect neutrophil migration and activation, which are key components of the inflammatory response. It is also believed to affect lymphocytes, but the precise mechanisms are complex and varied. It also influences endothelial cells, reducing vascular permeability and thereby limiting the leakage of fluids and immune cells into tissues (which causes swelling). Furthermore, salicylic acid modulates the expression of adhesion molecules on endothelial cells, which reduces the ability of immune cells to stick to and migrate through blood vessel walls. Oral administration of salicylate glucosides from wintergreen oil relieves abdominal pain and contractions by moderating the inflammatory response without the ulcerogenic effects common to traditional salicylates (e.g., aspirin). Since lipophilic molecules tend to be better absorbed in the GI tract due the lipid-rich membrane of the intestinal epithelium, it can be hypothesized that ingestion of wintergreen oil may have similar or superior effects.

2. Analgesic Properties

Molecular or Cellular Target/Pathway: Central nervous system (CNS) pathways involved in pain signaling and counterirritant effects.

Mechanism: Topical application of wintergreen essential oil as a methyl salicylate ointment alleviates headache symptoms by modulating CNS pain pathways and enhancing blood circulation in the localized area. When applied topically, wintergreen causes mild irritation and increases localized blood flow (counterirritant). This increased blood flow creates a warming sensation, which can help to distract from deeper pain signals, such as musculoskeletal pain.

3. Antioxidant Activity

Molecular or Cellular Target/Pathway: Reactive oxygen species (ROS), lipid molecules.

Mechanism: Wintergreen essential oil demonstrates significant antioxidant properties by scavenging free radicals, reducing lipid peroxidation, and providing protective measures against oxidative stress in cells.

4. Antimicrobial Effects

Molecular or Cellular Target/Pathway: Pathogen cell walls/membranes.

Mechanism: The components of wintergreen oil disrupt the cell membranes of bacteria and inhibit growth, demonstrating effectiveness against both gram-positive and gram-negative bacterial strains as well as fungal pathogens.

5. Insecticidal Activity

Molecular or Cellular Target/Pathway: Insect nervous systems.

Mechanism: Wintergreen essential oil acts as a biocide by delivering toxic effects to pests like the granary weevil and spotted wing drosophila, likely through neurotoxic pathways that lead to paralysis and death.

6. Antileishmanial (Antiparasitic) Activity

Molecular or Cellular Target/Pathway: Parasitic protozoa of the *Leishmania* genus.

Mechanism: Wintergreen essential oil exhibits antileishmanial activities, potentially affecting the viability of the protozoa through mechanisms that disrupt metabolic and reproductive functions, offering a promising avenue for treating cutaneous leishmaniasis.

YLANG YLANG

Derived from the exotic flowers of the *Cananga odorata* plant, ylang ylang essential oil is acclaimed for its wide-ranging therapeutic benefits, including mood enhancement, anxiety reduction, blood pressure modulation, and antimicrobial properties. The oil is often separated into different fractions, which are like different layers of scents, each with its own unique smell and properties. The main fractions include ylang ylang I, II, III, and extra (IV). The first fraction (ylang ylang I) has the strongest and sweetest aroma, while the later fractions are richer and more complex, with deeper notes. Each fraction can be used for different purposes in perfumes and aromatherapy, allowing for a variety of experiences and benefits, such as promoting relaxation or enhancing mood. Ylang ylang complete combines the different fractions of the oil, typically including all or a significant portion of the I, II, and III fractions, and sometimes even the IV fraction. This blend is often referred to as the "complete" version because it encompasses the full spectrum of ylang

ylang's fragrance. Research reveals that ylang ylang oil acts through various molecular mechanisms involving neurotransmitter modulation, antioxidant action, and antimicrobial activity, positioning it as a versatile agent in promoting physical and psychological health. Its components, such as linalool, benzyl benzoate, and geranyl acetate, contribute to its efficacy in treating conditions like hypertension, anxiety, pain, and even certain skin disorders. Furthermore, the oil shows promise in reducing the impacts of alcohol-related liver damage and supporting wound healing.

SUGGESTED THERAPEUTIC USE:

Cardiovascular Issues: Essential hypertension (high blood pressure), stress-induced cardiovascular strain

Mental Health Issues: Anxiety, low self-esteem, stress, mood disorders

Organ and Tissue Protection: Oxidative stress (cancer, IBD, diabetes, neurodegenerative conditions, cardiovascular disease, liver or kidney disease, metabolic syndrome, fibromyalgia, chronic fatigue syndrome, skin aging), cellular damage

Infections: Bacterial infections, fungal infections, toxin production by pathogens

Inflammatory Conditions: Acute inflammatory conditions, arthritis

Neurological Issues: Symptoms related to autism spectrum disorder (ASD), cognitive and social function impairment

Insect Control and Repellency: Mosquito and tick infestations

Liver Issues: Alcohol-induced liver damage, liver dysfunction

MOLECULAR AND CELLULAR TARGETS, BIOLOGICAL PATHWAYS, AND MECHANISMS:

1. Cardiovascular Effects

Molecular or Cellular Target/Pathway: Cortisol levels, sympathetic nervous system pathways.

Mechanism: Inhalation of ylang ylang oil reduces cortisol levels and modulates heart rate and blood pressure in individuals with essential hypertension, suggesting that its calming effects can help alleviate stress-induced cardiovascular strain.

2. Anxiolytic Properties

Molecular or Cellular Target/Pathway: Neurotransmitter receptors (serotonin, dopamine).

Mechanism: Animal studies indicate that ylang ylang essential oil modifies neurotransmitter levels, decreasing dopamine and increasing serotonin, leading to reduced anxiety, particularly notable in male subjects. Isolated benzyl benzoate affected levels of dopamine and serotonin receptors in a manner similar to ylang ylang.

3. Antioxidant Activity

Molecular or Cellular Target/Pathway: Reactive oxygen species (ROS) and free radicals.

Mechanism: Ylang ylang essential oil exhibits significant antioxidant properties, demonstrating its ability to scavenge free radicals and reduce lipid peroxidation, which contributes to cell preservation and overall health.

4. Antimicrobial Activity

Molecular or Cellular Target/Pathway: Pathogen virulence and growth factors, and their production of toxins.

Mechanism: *In vitro* studies show that ylang ylang oil prevents bacterial adherence and inhibits fungal growth, contributing to its role in infection control, particularly in clinical and agricultural settings. It also suppresses the production of aflatoxin B1 by fungi.

5. Anti-inflammatory Properties

Molecular or Cellular Target/Pathway: Leukocytes and neutrophils, proinflammatory cytokines (IL-6).

Mechanism: Ylang ylang essential oil modulates leukocyte recruitment—reducing their movement into inflamed joint areas, decreasing joint swelling and reducing the amount of pain and discomfort—and reduces inflammation in models of acute inflammation, indicating its effectiveness in treating inflammatory conditions like arthritis. It also reduces the movement of certain immune cells, called neutrophils, to sites of inflammation, and their ability to engulf foreign particles, a process known as phagocytosis. Additionally, it can lower the levels of IL-6, a protein that promotes inflammation, and prevent damage to cartilage in joints.

By targeting these processes, ylang ylang oil can alleviate symptoms like swelling, pain, and joint damage associated with inflammation.

6. Neuroprotective Effects

Molecular or Cellular Target/Pathway: Serotonin and dopamine metabolism in the prefrontal cortex.

Mechanism: Inhalation of ylang ylang enhances social interactions and performance in tasks by improving anxiety and serotonin metabolism in preclinical models of autism spectrum disorder (ASD). The relationship between these neurotransmitters and ASD is still being researched, but dopamine system variations in individuals with ASD could affect motivation to engage in social interactions, repetitive behaviors, emotional regulation and motor coordination. Similarly, imbalances in serotonin levels or the way it is transported to and utilized in the brain may contribute to sensory sensitivities, repetitive behaviors, and challenges with social interaction.

7. Effects on Mood and Self-esteem

Molecular or Cellular Target/Pathway: Neurotransmitter signaling pathways related to mood.

Mechanism: Both topical application and inhalation of ylang ylang oil have been shown to elevate self-esteem and promote a relaxed state among individuals, highlighting its mental health benefits. Self-esteem is a complex psychological construct influenced by various biological, psychological, and social factors. While it is considered primarily as a psychological response, some biological pathways and molecular targets have been implicated in the regulation of self-esteem. This includes neurotransmitters (serotonin, dopamine, norepinephrine), hormones (cortisol, oxytocin), genetics (SLC6A4, DRD2, DRD4, Val66Met, OXTR), epigenetics, brain function (prefrontal cortex, amygdala, hippocampus), neuroinflammation, and neuroplasticity. Preclinical research suggests that ylang ylang does affect neurotransmitters like dopamine and serotonin, and clinical research shows cortisol-lowering effects. The oil also influences the function of the prefrontal cortex in preclinical models. Altogether, the research suggests that ylang ylang most likely influences self-esteem through various mechanisms.

8. Insect Repellent and Acaricidal Activity

Molecular or Cellular Target/Pathway: Specific receptors in insect nervous systems.

Mechanism: Ylang ylang essential oil exhibits substantial repellent activity against mosquitoes and tick species, impairing their fertility and interrupting their nervous system function, making it a promising natural alternative for pest control.

9. Liver Health Support

Molecular or Cellular Target/Pathway: Liver function biomarkers (e.g., AST, ALT), inflammatory pathways (e.g., TNF, NF-kappa B).

Mechanism: Oral administration of ylang ylang oil significantly improves liver function after alcohol-induced liver damage by regulating proteins associated with liver toxicity and supporting antioxidant enzyme activity. Liver enzymes are substances produced by your liver that help with important processes in your body—like breaking down food and detoxifying harmful substances. When doctors want to check how well your liver is functioning and its overall health, they often measure levels of certain liver enzymes (ALT, AST, ALP, GGT) in your blood. When your liver is healthy, it produces and releases enzymes into your blood at normal levels. If the liver is injured or diseased (due to things like fatty liver, hepatitis, or alcohol abuse), the liver cells can get damaged or inflamed. This can cause more enzymes to leak into the bloodstream. Liver enzymes act like warning signs—normal enzyme levels suggest a healthy liver, while elevated levels can indicate liver trouble.

Best Practices for the Best Results

First and foremost, you must use a pure, unadulterated, high-quality essential oil for therapeutic purposes! Doing so is crucial to achieve consistent and reliable results that users depend on for health and well-being. Essential oils are highly concentrated extracts from plants, and their potency can significantly impact their effectiveness in promoting physical, emotional, and mental health. Indeed, essential oil chemistry directly affects safety and efficacy. You can't expect a peppermint essential oil with only five percent menthol to provide the same results as one with forty percent menthol. Nor can you expect a lavender with synthetic linalyl acetate to produce the same effect as a lavender oil with natural linalyl acetate because synthetic molecules affect the body differently. When essential oils are pure, they contain the full spectrum of beneficial compounds, ensuring that users experience the intended health benefits, whether for relaxation, pain relief, or other therapeutic effects.

Unfortunately, the market for essential oils is rife with significant adulteration, with estimates suggesting that up to eighty percent of essential oils sold are either contaminated, diluted, synthesized with artificial ingredients, or poor quality. This not only diminishes their therapeutic efficacy but can also pose safety risks. Adulterated oils may cause adverse reactions or fail to provide the desired therapeutic effects, leading users to believe that essential oils are ineffective or harmful. Thus, it is essential for consumers to source their essential oils from reputable suppliers who ensure purity and quality through rigorous testing and transparent sourcing practices. Prioritizing pure essential oils is vital for both safety and achieving the potential health benefits these natural substances can offer.

Below are best practices based on research and clinical experience. Remember, these are guidelines and need to be adjusted to your personal situation, including health status, age, sensitivities, size, and more.

INHALATION

Inhalation is a safe and easy way to enjoy the benefits of essential oils. The aromatic benefits of essential oils are rooted in neuroscience because the sense of smell is directly linked to the brain. Smells are processed first by the olfactory bulb, which starts in the nose and runs along the bottom of the brain. The olfactory bulb shuttles odor molecules to areas of the brain it has a direct connection to—like those found in the limbic system. The limbic system includes the amygdala, which is the part of the brain involved in personal experiences and emotions; and the hippocampus, which is involved in memory and associative learning, so influencing this system has a great impact on mood, stress, memory, learning, and anxiety. The limbic system also influences endocrine and nervous system function, sleep, libido, appetite, heart rate and blood pressure, and hormone balance, so it is easy to see how inhalation of essential oils can be a powerful experience. In fact, the limbic system initiates responses to the presence of odor molecules that can influence pain, metabolism, and your overall well-being.

When you inhale essential oils, their aromatic molecules travel through your nose and into your lungs. Inside the nose, these molecules interact with special nerve receptors in the olfactory system, which send signals directly to the brain's limbic system—the area responsible for emotions, memory, and certain physiological responses like heart rate and stress levels. Meanwhile, as you breathe in, the essential oil molecules also reach the lungs, where they pass through the thin walls of the alveoli (tiny air sacs) and enter the bloodstream. From there, they circulate to various organs and tissues in the body. This dual pathway—directly to the brain through smell and into the bloodstream via the lungs—explains why inhaling essential oils can have both immediate mental and emotional effects as well as broader physiological benefits.

Inhalation of essential oils leads to rapid absorption and distribution to the bloodstream due to the efficient exchange that occurs at the alveoli. Essential oil compounds are small and can easily pass through the thin walls of the alveoli and enter the capillaries (the small blood vessels). Once in the bloodstream, these compounds are quickly distributed throughout the body, allowing users to feel their effects almost immediately. This rapid absorption makes inhalation an effective method for using essential oils, especially for promoting relaxation, enhancing mood, or alleviating stress, as the compounds can reach the brain and other organs quickly. Indeed, aroma compounds are detected in the blood within minutes, peaking around twenty minutes after inhalation.[91] A significant portion of essential oil compounds are absorbed and distributed to the pulmonary system following inhalation.[92,93] The research showed that the monoterpene limonene was processed relatively quickly, but some of it built up in fatty tissues. Animal research suggests that inhalation of aroma compounds leads to distribution in the brain tissue.[94] So, inhalation, although simple, leads to distribution of essential oils primarily to the lungs and secondarily to the bloodstream and brain.

Best practices for inhalation include:

- Inhale directly from the bottle—Simply remove the bottle cap and inhale (Note:

this will expose you to less than the full complex of the oil as the lighter molecules will escape from the bottle first)

- Diffusers
  - ☐ Water-based ultrasonic diffusers
    - ➤ These use ultrasonic vibrations (very high-frequency sound waves) to create a fine mist
    - ➤ Pros: Quiet, versatile, humidifying, economical
    - ➤ Cons: Creates an oil-water mixture of mist with a lower concentration than other diffusers
    - ➤ Add 1–4 drops of EO per 100 mL of water
  - ☐ Nebulizing diffusers
    - ➤ Leverage pressurized air to break down the essential oil into a very fine mist
    - ➤ Pros: Strongest aroma, pure EO diffusion, more therapeutic
    - ➤ Cons: More expensive, typically louder, uses EO more quickly
    - ➤ Direct connection of pure oil to unit.
  - ☐ Evaporative diffusers
    - ➤ Uses a fan to blow air through a pad or filter that's been soaked with essential oil
    - ➤ Pros: Simple, inexpensive, portable
    - ➤ Cons: Aroma can be inconsistent, lighter constituents evaporate first
    - ➤ Add 3–5 drops of oil to the pad as a starting point and adjust from there as needed
  - ☐ Heat diffusers
    - ➤ Heat to warm the essential oil, causing it to evaporate into the air
    - ➤ Pros: Inexpensive
    - ➤ Cons: Heat can alter the chemical composition of the EO, potentially affecting its therapeutic benefits, potential fire hazard
    - ➤ Start with 23 drops and adjust depending on desired strength of aroma.
  - ☐ Passive diffuser (stone, terracotta, etc.)
    - ➤ Porous materials like terracotta and certain stones readily absorb Eos, and it slowly evaporates into the surrounding air
    - ➤ Pros: Inexpensive, easy to use, portable
    - ➤ Cons: Limited diffusion range, less intense aroma, slower diffusion, aroma can be inconsistent, oil waste (some of the oil remains in the passive diffuser)
    - ➤ Add 3–5 drops to the surface
  - ☐ Personal inhaler diffuser
    - ➤ Small portable devices, often resembling a tube, that contain a cotton wick or similar absorbent material that EOs are applied to and then inserted into the nose for inhalation

- ➤ Pros: Inexpensive, portable, concentrated EOs delivered through the nostril, similar to inhaling directly from the bottle
- ➤ Cons: Aroma can be inconsistent, lighter constituents evaporate first
- ➤ Add 10–20 drops of EO to the cotton wick (4–8 drops for children aged 5+)
- Steam Inhalation—add 1 to 2 drops of essential oil in a bowl of steamy water (not boiling), close your eyes, cover your head and the bowl with a towel and lean over the bowl, inhale deeply through your nose and exhale through your mouth, continue this process for three to five minutes and repeat several times throughout the day
- Palm diffusing—Add 1–2 drops of EO to one palm, rub palms together, close your eyes, cup your hands over your nose and mouth and breathe deeply for a few minutes
- Jewelry—add 3 to 5 drops of essential oil to the stone or cotton and insert into jewelry, then lift the jewelry to your nose and inhale as needed

TOPICAL

When you apply essential oils to your skin, their small molecules penetrate through the outermost layer (epidermis) and reach the deeper layers of skin and underlying tissues. Some of the oil stays in the area where it was applied, interacting with local cells and tissues to provide targeted benefits like reducing inflammation, soothing sore muscles, or promoting healing. Meanwhile, some of the essential oil molecules pass through the skin's lipid barrier and enter tiny blood vessels (capillaries) in the dermis. From there, they get absorbed into the bloodstream and circulate in the body, reaching different organs and systems. Unlike inhalation, which leads to quicker systemic distribution, topical absorption allows for a slower release, with some essential oil molecules being temporarily stored in fat cells or muscle tissue. These stored molecules are gradually released over time, extending the effects of the essential oil beyond the initial application. This combination of localized action and gradual systemic absorption makes topical application an effective way to experience both immediate and lasting benefits.

Research suggests that essential oil compounds, such as those found in lavender, can be detected in the blood within just five minutes of topical application, with peak blood concentration occurring around 19 minutes and declining to a minimum after approximately 90 minutes.[95] The absorption and distribution of these compounds depend on their chemical structure. Oxygenated monoterpenes—such as linalool, linalyl acetate, 4-terpineol, menthol, 1,8-cineole, and bornyl acetate—are well-absorbed, whereas hydrocarbons like limonene, alpha- and beta-pinene, chamazulene, and beta-caryophyllene are absorbed less efficiently. One study found that only trace amounts of a 5% dilution of tea tree essential oil remained in the skin layers, with oxygenated monoterpenes being significantly absorbed while hydrocarbons were present only in minimal quantities.[96] These findings highlight the importance of an essential oil's chemical makeup in determining its effectiveness and bioavailability after topical application.

Best practices for topical application involve:

- Proper dilution before topical application reduces sensitization, irritation, unpleasant reactions, and improves efficacy
  - ☐ Some falsely believe that applying neat (without carrier oil) is more effective, but the reality is less of the oil stays in contact with the skin and more evaporates into the air, decreasing efficacy. Dilution reduces potency, but not efficacy
  - ☐ For simplification, one drop of EO in 5 mL (a teaspoon) of carrier oil equals a 1% dilution; so to get a 10% dilution you need to add 10 drops EO to 5 mL of carrier oil
  - ☐ General dilution recommendations:
    - ❖ General use—4%–10%
    - ❖ Facial use—0.5%–3%
    - ❖ Body massage—3%–5%
    - ❖ Pain—7.5%–50%
    - ❖ Digestive (abdominal)—5%–17.5%
    - ❖ Skin issues—1%–25%
    - ❖ Acute situations—5%–15%
    - ❖ Chronic situations—5%–10%
    - ❖ Bug bite, canker sore, toenail fungus—neat (only occasional use)
    - ❖ Specific EOs will need to be diluted more

- 4/10/75 guideline—About 4% of an EO is absorbed when applied neat, 10% when diluted, and it increases to 75% when the application is covered (semi-occlusive)
- Start small, 1–2 drops of EO, and apply more frequently
- Dilute more for children, elderly, people with sensitive skin or compromised immune system
- Dilute irritating (called "hot" oils) like cassia, cinnamon, thyme, oregano, cumin, and lemongrass
- Some EOs—lemon, grapefruit, bergamot, lime, etc. contain furanocoumarins which can increase sensitivity to UV light—avoid UV rays for at least 12 hours after topical application of these oils
- Do not apply EOs directly in the ears or eyes
- Dilute more when applying on or around sensitive tissues (e.g., the face or genitals)

INGESTION

Humans have been ingesting small amounts of compounds found in foods, herbs, and spices for centuries. Citrus fruits provide limonene and other terpenes, herbs and spices contain eugenol, thymol, carvacrol, menthol, and more. Humans metabolize essential oil compounds through a variety of enzymatic processes primarily in the liver. These compounds exert various biological effects that help maintain and restore health. The metabolism of these compounds often involves phase I and phase II reactions, including oxidation and conjugation, which help make these compounds more water-soluble for excretion.

Essential oils are complex mixtures of volatile aromatic compounds. These compounds can be broadly classified into two main categories based on their chemical structure: oxygenated and non-oxygenated. Each essential oil can contain from a dozen to hundreds of different compounds, though some are present in trace amounts.

Non-oxygenated compounds: These are primarily hydrocarbons, meaning they consist of carbon and hydrogen atoms. The absorption of non-oxygenated compounds can be slower compared to oxygenated compounds because of their lower polarity—less attracted to water, more oil-soluble. These lipophilic compounds often require dietary fats to facilitate their absorption. These compounds tend to have a longer half-life (time for half of a drug to be eliminated from the body) in systemic circulation, which can be advantageous for therapeutic effects. They include:

- Terpenes: These are the most abundant components in many essential oils. They are built from isoprene units (a 5-carbon molecule). Examples include:
  - Monoterpenes: (10-carbon) like limonene (citrus scent), alpha-pinene (pine scent), beta-pinene, myrcene.
  - Sesquiterpenes: (15-carbon) like beta-caryophyllene (found in many spices), alpha-humulene.
- Aromatic Hydrocarbons: These contain a benzene ring. Examples include:
  - Toluene: This is a simple aromatic hydrocarbon with a benzene ring and a methyl group. It's not commonly a major component of essential oils but can be present in trace amounts in some.
  - Xylenes: These are also aromatic hydrocarbons with a benzene ring and two methyl groups. Like toluene, they are not typical major components of essential oils but may be present in trace amounts.

Oxygenated compounds: These contain oxygen in addition to carbon and hydrogen. They tend to be more polar (dissolve better in polar solvents like water or hydrosols) and often contribute significantly to the aroma of essential oils. Increased polarity often enhances their absorption in the gut. Compounds like linalool (an alcohol) and menthol (a terpene) are examples of oxygenated compounds that are rapidly absorbed due to their favorable solubility profile. They include:

- Alcohols: Like linalool (lavender scent), geraniol (rose scent), menthol (peppermint scent), alpha-terpineol.
- Aldehydes: Like citral (lemon scent), citronellal (citronella scent).
- Ketones: Like camphor, menthone (peppermint scent), carvone (spearmint scent).
- Acids: Like benzoic acid.
- Esters: Like ethyl acetate, linalyl acetate (lavender scent).
- Phenols: Like eugenol (clove scent), thymol (thyme scent).
- Ethers and Oxides: Like 1,8-cineole (eucalyptus scent).

- Coumarins: Some coumarins can be found in essential oils and have a characteristic sweet, hay-like aroma.

The oral bioavailability of essential oil compounds varies significantly based on their chemical structure, lipophilicity, and molecular size.

Monoterpenes (e.g., limonene, thymol, linalool):

- Characteristics: Typically have low molecular weights and are lipophilic.
- Absorption: Rapidly absorbed through the gastrointestinal tract due to their lipophilicity.
- Bioavailability: Studies suggest that monoterpenes are quickly absorbed after oral administration, but specific bioavailability percentages can vary.[97]

Sesquiterpenes (e.g., β-Caryophyllene, α-Zingiberene, patchoulol):

- Characteristics: Larger molecular structures compared to monoterpenes and are also lipophilic.
- Absorption: Their larger size may result in slower absorption rates.
- Bioavailability: Limited specific data is available, but their lipophilic nature suggests they are absorbed in the small intestine along with dietary lipids.[98]

Diterpenes (e.g., sclareol, incensole, incensyl acetate):

- Characteristics: Even larger molecular structures and are highly lipophilic.
- Absorption: Due to their size and lipophilicity, they may have lower absorption rates and bioavailability.

- Bioavailability: Specific data is scarce, but their absorption is likely limited without appropriate delivery systems due to their poor water solubility, rapid excretion, and slow dissolution rates.[99]

Several factors influence bioavailability, such as:

Lipophilicity: Highly lipophilic compounds tend to form micelles (spherical molecules that have one end that loves water and one end that hates water arranged with the water-loving ends on the outside and the water-hating ends tucked inside, creating a little pocket to carry things that don't like water, like grease or oil, through the water) and are digested in the small intestine along with other lipids.[100]

Molecular Size: Larger molecules may have reduced absorption rates due to their size.

Metabolism: Compounds may undergo first-pass metabolism in the liver, reducing their systemic availability.

Pharmacokinetics studies what the body does to a substance. Think of it like the journey of the substance through the body: how it's absorbed, distributed, broken down, and eliminated. On the other hand, pharmacodynamics studies what the substance does to the body. Think of it like the substance's effect on the body: how it interacts with cells and tissues to produce a therapeutic effect. Understanding pharmacokinetics and pharmacodynamics is crucial for the safe and effective use of therapeutics. It helps determine the right dose to produce therapeutic effects, optimize delivery, understand tissue distribution, minimizes risks by defining a therapeutic window (the amount in a range that will produce a positive result without adverse effects), and guides personalized treatments. In essence, pharmacokinetics and pharmacodynamics provide the scientific foundation for using therapeutics safely and effectively. They help us understand how to get the right amount of the right therapeutic to the right place in the body at the right time to achieve the desired therapeutic effect while minimizing risks.

Several clinical studies have investigated the pharmacokinetics of D-limonene, a natural monoterpene found in citrus oils, to understand its absorption, distribution, metabolism, and excretion in humans. The research found that the limonene was well-tolerated even at high doses.[101] Limonene was rapidly absorbed and preferentially concentrated in breast tissue.[102, 103] These studies collectively indicate that D-limonene is well-absorbed when administered orally, undergoes metabolism to active compounds like perillic acid, and preferentially accumulates in certain tissues.

Pharmacokinetic and pharmacodynamic research is ongoing and will help guide the creation of more effective formulas with the right amount of essential oil. This work is crucial for the therapeutic use of essential oils to advance them into the future.

To improve the bioavailability of these compounds, various strategies can be employed:

Adding fatty oils to softgels or capsules. Adding fatty oils like coconut or olive oil— or encapsulating them in lipids, like liposomes—can enhance the oral bioavailability of essential oils by improving their solubility and absorption in the gastrointestinal tract, allowing for better distribution to the bloodstream.[104] Make at least half of your capsule a fatty oil.

Adding essential oils to an herbal powder. Incorporating essential oils into a powder reduces its bloodstream absorption and increases its absorption and retention in the gastrointestinal (GI) tract, potentially benefiting the treatment of irritable bowel syndrome, by forming complexes that are more easily absorbed and retained in the GI mucosa.[105]

Sublingual or buccal cavity absorption. Sublingual or buccal absorption bypasses the harsh environment of the digestive system and first-pass metabolism in the liver, allowing more of the substance to enter the bloodstream directly and potentially increasing its bioavailability compared to swallowing a capsule.[106] Substances that undergo high first-pass metabolism, like essential oils, generally have significantly higher bioavailability when used sublingually.

Advanced Delivery Systems. Formulations such as self-emulsifying drug delivery systems (SEDDS) have been shown to significantly increase the oral bioavailability of certain essential oil compounds.[107, 108]

Understanding these factors is crucial for optimizing the therapeutic efficacy of essential oil compounds when administered orally. As a best practice, you should take your essential oils with a fatty substance to improve bioavailability.

In conclusion, the oral bioavailability of essential oil constituents is a complex interplay of chemical structure, lipophilicity, and molecular size, all of which significantly influence their absorption and therapeutic efficacy. With a growing body of research shedding light on the pharmacokinetics and pharmacodynamics of these aromatic compounds, it becomes increasingly clear that the methods of administration and enhancement play a pivotal role in maximizing their bioactive potential. From innovative delivery systems to the strategic incorporation of fatty oils, these advancements not only help improve absorption rates but also pave the way for personalized therapeutic applications. As we continue to explore the vast landscape of essential oils, understanding and utilizing their bioavailability will be essential for harnessing their full therapeutic promise and ensuring their safe and effective use in our daily lives. By embracing these findings, we can unlock the true potential of essential oils, transforming them into powerful allies in holistic health and wellness.

Ideally, all essential oil dosing would be determined based on pharmacokinetic and pharmacodynamic data along with clinical research. Some essential oils have a well-established therapeutic dose—80 mg (1.8 drops) of lavender for anxiety, twice daily; and 0.1–0.2 mL (2.25–4.5 drops) of peppermint, three times daily for IBS—while others have dosing recommended by organizations like the European Medicines Agency or the German Commission E.

Some of these recommendations are as follows:

- Caraway—0.15–0.3 mL/day; 3–6 drops/day.[109, 110]
- Cinnamon—50–200 mg/day.[111]
- Cornmint (Japanese mint; *Mentha arvensis, M. canadensis*)—3–6 drops/day.[112]
- Eucalyptus (1,8-cineole rich *Eucalyptus* species)—0.3–0.6 mL/day; 100–200 mg, 2–5 times per day.[113, 114]
- Fennel—0.1–0.6 mL/day.[115]
- Niaouli—200–2,000 mg/day.[116]
- Thyme—6–12 drops/day.[117]

Best practices for ingestion (require adjustment based on individuality and the essential oil(s) being ingested):

- Take 1 to 5 drops in a capsule every 4 to 6 hours
  - ☐ Enteric or pH-dependent capsules can help deliver the oils to the small intestine where they are designed to be absorbed
  - ☐ Fill at least half of the capsule with a fatty oil (medium-chain triglycerides, olive oil, avocado oil, etc.)
- Take a total of 25 drops in a 24-hour period
- 1–2 drops sublingually or on the tongue, holding in mouth, 3 to 5 times daily (Note: Sublingual delivery can provide a 14.5-fold increase in bioavailability, while oromucosal delivery can increase bioavailability 15.5-fold.).[118]
- Mild oils can be added to water, tea, milk, or other beverages (Note: Due to the hydrophobic nature of essential oils, this forms droplets that are harder to

absorb into the bloodstream; additionally, the oils are exposed to stomach acid, which breaks down the compounds, decreasing bioavailability and dispersion to tissues)

- Take essential oils internally with food and at least eight ounces of water
- Add to dishes (only use small amounts, such as a toothpick dip or one drop depending on the dish)

The way you use an essential oil can be guided by the system or body part you are trying to affect. Their effective use can be greatly enhanced by understanding the body systems or specific body parts that you aim to support or influence. Each essential oil has unique properties and therapeutic attributes that can target various physical, emotional, and mental health concerns. Understanding the "best" way to use an essential oil for a specific body system allows for a more targeted and effective approach to wellness. The preceding chapters provide information that can guide your selection of the "right" essential oils for your concern, while the information below can help identify the appropriate application methods—such as inhalation, topical application, or ingestion—to enhance your therapeutic experiences and results. Always consider personal sensitivities and consult with a qualified healthcare provider or aromatherapist when integrating essential oils into your health regimen, particularly for therapeutic purposes.

Integumentary System. This system comprises the skin, hair, and nails and serves to protect the body from external threats, regulate body temperature, and provide sensory information. Most of the time when addressing issues in the integumentary system, you will want to apply the oil diluted to the affected area. The exception to this is when a skin issue like eczema or psoriasis is involved. These conditions are usually largely gut-driven, meaning you will need to support gut integrity and the gut microbiome by ingesting essential oils and other nutraceuticals.

Skeletal System. Consisting of your bones, cartilage, and ligaments, the skeletal system provides structural support, protects organs, enables movement, and stores minerals. Again, you will likely be applying essential oils to the affected area—knee for knee pain, shoulder for shoulder pain, etc.—when addressing skeletal issues. However, some of the molecular targets may require ingestion of an oil as well. For instance, if you have osteoporosis, which is a condition characterized by imbalanced osteoclast (break down bone) to osteoblast (build new bone), you will likely apply topically and perhaps ingest an oil as well. Additionally, if you need support maintaining a healthy inflammatory response systemically, ingestion would be a good idea. Hormones like estrogen and parathyroid hormone play significant roles in bone metabolism as well, so you might inhale or ingest an oil to influence hormone balance.

Muscular System. The muscular system includes skeletal, smooth, and cardiac muscles, which facilitate body movement, maintain posture, and generate heat. More recently, researchers have discovered that muscles produce molecules called myokines that influence both mood and brain function. When you exercise, your muscles release these signaling proteins into the bloodstream, where they interact with the brain and nervous system, impacting mood, stress resilience, and mental health. These molecules include BDNF (cognitive health and mood), irisin (reduced anxiety and depression), IL-6 (regulates inflammation, improves mood and mental clarity), cathepsin B (cognition), lactate (mood), and modulates the kynurenine pathway (stress resilience, mood).

Nervous System. Made up of the brain, spinal cord, and nerves, this system controls and coordinates bodily functions, processes sensory information, and facilitates cognition. The fastest way to influence the central nervous system is through inhalation, which rapidly delivers essential oil molecules across the blood-brain barrier and into the brain. The effects of inhaled essential oils can be felt relatively quickly because of this. It is also possible to deliver essential oils to the nervous system via topical application to the spine and head. Since peripheral nerves are located throughout the body, you may also apply essential oils to a local area, such as on the wrist when experiencing carpal tunnel syndrome. Lastly, when ingested, essential oils are absorbed through the digestive system and then enter the bloodstream, where they are distributed throughout the body, including the nervous system.

Immune System. The immune system involves a very complex network of white blood cells, organs, mucous membranes, bone marrow, and antibodies that act as the body's defense network against harmful invaders like bacteria, viruses, fungi, and parasites. It also targets abnormal cells, such as cancer cells. It has crossover with the skin and lymphatic system. Most of the time you will ingest an essential oil to affect immune system activity. By ingesting an essential oil, the constituents can interact with cells involved in immune responses. However, during periods of acute illness, it is helpful to inhale essential oils as well to support healthy respiratory function.

Cardiovascular System. Consisting of the heart, blood vessels, and blood, the cardiovascular (also circulatory) system transports oxygen, nutrients, and waste products throughout the body. For the cardiovascular system, getting essential oils into the bloodstream at sufficient levels is important. This likely means that you will need to ingest the oil. Then again, if your intent is to decrease blood pressure, inhalation has been a proven method in multiple clinical studies and tends to have swift effects. You can also consider applying the essential oils over the heart area of the chest and large muscles where capillaries (the entry point for essential oils after topical application) are densely concentrated. Regarding topical application, it is highly improbable that essential oils would directly and specifically enter large arteries after topical application. Rather, they would likely enter systemic circulation once reaching the capillaries in the dermal layer of the skin.

Lymphatic System. Often overlooked, the lymphatic system is a series of vessels, lymph nodes, and organs like the spleen that supports the immune system by filtering fluids and fighting infections. The lymph system requires movement or massaging of tissues for the lymph fluid to circulate since it does not have a pump like the heart. Moving or massaging large muscles can effectively move lymph. To get essential oils into the lymph system, they first need to enter systemic circulation. The lymphatic system works closely with the cardiovascular system, making it plausible that essential oils in the bloodstream can enter lymph fluid. Especially since essential oils are lipophilic, which may make it easier for them to enter the lymphatic system, since the lymph system carries fats. Ingestion may be the best way to deliver the highest concentration of essential oils to the lymph system, but this may also be accomplished to a lesser degree through topical application—especially when applied to large muscles that are then massaged or exercised.

Respiratory System. Consisting of your lungs, trachea, and bronchi, the respiratory system facilitates gas exchange (oxygen and carbon dioxide) between the body and the environment. To affect the lungs, you'll want to primarily focus on inhalation. Essential oils are of the size and weight that they can get to both the upper and lower respiratory tract, providing far-reaching effects in the respiratory system. The lungs have an extremely dense capillary network surrounding the alveoli, where gas exchange occurs, suggesting inhalation will deliver essential oils into the bloodstream as well.

Digestive System. This system that includes the mouth, stomach, intestines, liver, and gallbladder, breaks down food, absorbs nutrients, and eliminates waste. The digestive system is influenced by all three primary administration methods. Ingestion of oils gets the oils into the digestive tract where they can support the microbiome, gut integrity, fight pathogens, and relax the muscles of the GI tract. Inhalation is a fast way to reduce nausea and vomiting, while topical application, particularly in a clockwise circular motion to the abdomen, supports healthy GI motility and bowel regularity.

Urinary System. Comprising the kidneys, bladder, and ureters, the urinary system filters waste products from the blood and regulates fluid balance. Your kidneys are also highly involved in blood pressure regulation. When your blood pressure is too high, the kidneys remove extra water and salt from your blood, which then leaves your body as urine, lowering the blood volume and therefore the pressure. If your blood pressure is too low, the kidneys hold onto water and salt, increasing blood volume and raising the pressure. Salt plays a big role because water follows salt; so, the more salt your kidneys hold onto, the more water they retain, and the higher your blood pressure can get. The primary way to influence the urinary system is by ingesting essential oils, but you can also apply them over the bladder or kidney areas.

Reproductive System. In males, the reproductive system includes the testes, penis, and prostate gland. The reproductive system is closely linked to the endocrine system, which regulates hormones, and the urinary system. In women, it includes the ovaries, uterus, and vagina. Regardless of sex, this system serves to facilitate sexual reproduction. However, there are nuances for its function in men and women. The reproductive system produces testosterone in men, which facilitates the development of male secondary sexual characteristics (e.g., facial hair, deepening of the voice), maintains muscle and bone mass, and triggers sex drive. The prostate gland secretes fluid that contributes to semen and plays a role in urinary function. The primary purpose of ovaries in females is to produce eggs (ova), estrogen, and progesterone, which develop female secondary sexual characteristics (e.g., breast development). They are also involved in regulation of the menstrual cycle, bone health, and mood. The uterus is involved in the menstrual cycle and is designed to house a developing fetus. Healthy reproductive function can be affected in all three ways, but ingestion is likely the most effective. Research shows that inhaling essential oil can balance hormone levels due to the neuroendocrine effects of essential oils traveling through the olfactory system. External genitalia will likely be most affected by topical application. Just keep in mind that these are very sensitive tissues and the most highly absorbable places on the human body, so dilute well and use less oil.

As a last point, you can apply essential oils to reflex points—on the feet, hands, ears, or meridian points—corresponding with the area of your body that needs support. To do so, apply the essential oil and then be sure to stimulate the point for long enough (generally thirty seconds to three minutes)

to produce a reflex response. A reflex response is the body's automatic reaction to the stimulation of specific reflex points on the feet, hands, or ears, that sends a message through the nervous system to promote healing and balance in corresponding areas of the body.

Incorporating essential oils into your wellness routine requires a thoughtful approach rooted in knowledge and best practices to maximize their benefits safely. By prioritizing pure, high-quality essential oils, you can avoid the pitfalls of adulteration that diminish their efficacy and pose potential risks. Understanding the various methods of application—whether through inhalation, topical use, or ingestion—allows for a tailored experience that aligns with your specific health goals, whether that be to promote relaxation, support physical well-being, or enhance emotional balance. Remember that individual factors, including your unique sensitivities and health status, should guide your choices and dosages. Staying informed of the latest research on the pharmacokinetics and pharmacodynamics of essential oils can further enhance your understanding of how these powerful natural compounds interact with your body. In doing so, you not only enrich your holistic health experience but also empower yourself to harness the full therapeutic potential of essential oils, making them a valuable ally in achieving optimal health and wellness. Don't delay, embrace the transformative possibilities these botanical treasures have to offer, and experience greater wellness naturally today.

References

[1] Santha S, Bommareddy A, Rule B, et al. Antineoplastic Effects of α-Santalol on Estrogen Receptor-Positive and Estrogen Receptor-Negative Breast Cancer Cells through Cell Cycle Arrest at G2/M Phase and Induction of Apoptosis. *PLoS One*. 2013 Feb 22;8(2):e56982.

[2] Lopez V, Nielsen B, Solas M, et al. Exploring Pharmacological Mechanisms of Lavender (Lavandula angustifolia) Essential Oil on Central Nervous System Targets. *Front Pharmacol*. 2017 May 19;8:280.

[3] dos Santos ERQ, Maia JGS, Fontes-Junior EA, et al. Linalool as a Therapeutic and Medicinal Tool in Depression Treatment: A Review. *Curr Neuropharmacol*. 2022 May 16;20(6):1073–1092.

[4] Shin YK, Seol GH, et al. Effects of linalyl acetate on oxidative stress, inflammation and endothelial dysfunction: can linalyl acetate prevent mild cognitive impairment? *Front Pharmacol*. 2023 Jul 28;14:1233977.

[5] Bavarsad BH, SBagheri S, Kourosh-Arami M, et al. Aromatherapy for the brain: Lavender's healing effect on epilepsy, depression, anxiety, migraine, and Alzheimer's disease: A review article. *Heliyon*. 2023 Jul 20;9(8):e18492.

[6] Zhang N, Zhang L, Feng L, et al. The anxiolytic effect of essential oil of Cananga odorata exposure on mice and determination of its major active constituents. *Phytomedicine*. 2016 Dec 15;23(14):1727-1734.

[7] Okugawa H, Ueda R, Matsumoto K, et al. Effects of sesquiterpenoids from "Oriental incenses" on acetic acid-induced writhing and D2 and 5-HT2A receptors in rat brain. *Phytomedicine*. 2000 Oct;7(5):417-22.

[8] Li W, Yao L, Mi S, et al. [The effect of borneol on level of HA and 5-HT in rat's hypothalamus]. *Zhong Yao Cai*. 2004 Dec;27(12):937-9.

[9] Wang ZJ, Heinbockel T. Essential Oils and Their Constituents Targeting the GABAergic System and Sodium Channels as Treatment of Neurological Diseases. *Molecules*. 2018;23:1061.

[10] Pinto EP, Mendes da Costa SO, D'Haese C, et al. Poly-ε-caprolactone nanocapsules loaded with copaiba essential oil reduce inflammation and pain in mice. *Int J Pharmaceutics*. 2023 Jul 25;642:123147.

[11] Baylac S, Racine P. Inhibition of 5-lipoxygenase by essential oils and other natural fragrant extracts. *Int J Aromatherapy*. 2003;13(2-3):138-42.

[12] Koh KJ, Pearce AL, Marshman G, et al. Tea tree oil reduces histamine-induced skin inflammation. *Br J Dermatol*. 2002 Dec;147(6):1212-7.

[13] Funk JL, Frye JB, Oyarzo JN, et al. Anti-Inflammatory Effects of the Essential Oils of Ginger (Zingiber officinale Roscoe) in Experimental Rheumatoid Arthritis. *PharmaNutrition*. Author manuscript; available in PMC: 2017 Jul 1.

[14] Johnson Scott A. *Medicinal Essential Oils: The Science and Practice of Evidence-based Essential Oil Therapy* (second edition). Scott A Johnson Professional Writing Services, LLC, 2023.

[15] Accessed from: https://www.theresaneoforthat.com/the-complete-list-of-orac-ratings-for-essential-oils/ on February 28, 2025.

[16] Carson CF, Hammer KA, Riley TV. Melaleuca alternifolia (Tea Tree) Oil: a Review of Antimicrobial and Other Medicinal Properties. *Clin Microbiol Rev*. 2006 Jan;19(1):50–62.

[17] Sienkiewicz M, Łysakowska M, Denys P, et al. The antimicrobial activity of thyme essential oil against multidrug resistant clinical bacterial strains. *Microb Drug Resist*. 2012 Apr;18(2):137-48.

[18] Johnson Scott A. *Medicinal Essential Oils: The Science and Practice of Evidence-based Essential Oil Therapy* (second edition). Scott A Johnson Professional Writing Services, LLC, 2023.

[19] Capatina L, Boiangiu RS, Dumitru G, et al. Rosmarinus officinalis Essential Oil Improves Scopolamine-Induced Neurobehavioral Changes via Restoration of Cholinergic Function and Brain Antioxidant Status in Zebrafish (Danio rerio). *Antioxidants (Basel)*. 2020 Jan 10;9(1):62.

[20] Saeedi M, Iraji A, Vahedi-Mazdabadi Y, Alizadeh A, et al. Cinnamomum verum J. Presl. Bark essential oil: in vitro investigation of anti-cholinesterase, anti-BACE1, and neuroprotective activity. *BMC Complement Med Ther*. 2022 Nov 18;22(1):303.

21 Chen WN, Chin KW, Tang KS, et al. Neuroprotective, Neurite Enhancing and Cholinesterase Inhibitory Effects of Lamiaceae Family Essential Oils in Alzheimer's Disease Model. HERMED-D-22-00403. Available at: https://ssrn.com/abstract=4112137 or http://dx.doi.org/10.2139/ssrn.4112137.

22 Al-Mijalli SH, Mrabti HN, Assaggaf H, et al. Chemical Profiling and Biological Activities of Pelargonium graveolens Essential Oils at Three Different Phenological Stages. *Plants (Basel)*. 2022 Aug 27;11(17):2226.

23 Gad HA, Mamadalieva RZ, Khalil N, et al. GC-MS Chemical Profiling, Biological Investigation of Three Salvia Species Growing in Uzbekistan. *Molecules*. 2022 Aug 23;27(17):5365.

24 Fraternale D, Flamini G, Ascrizzi R, et al. In Vitro Anticollagenase and Antielastase Activities of Essential Oil of Helichrysum italicum subsp. italicum (Roth) G. Don. *J Med Food*. 2019 Oct;22(10):1041-1046.

25 Koochi ZH, Jafromi KG, Kavoosi G, et al. Citrus peel waste essential oil: Chemical composition along with anti-amylase and anti-glucosidase potential. *Int J food Sci Tech*. 2022;57:6795-6804.

26 Sajid R, Abbas Z, Nazir M, et al. Valorization of hydro-distillate of fruit peels of Citrus paradisi macfad. Cultivar. Foster: Chemical profiling, antioxidant evaluation and in vitro and in silico enzyme inhibition studies. *Heliyon*. 2024 Aug 21;10(17):e36226.

27 Pacheco-Hernández Y, Rangel-Galván M, Velásquez-Hernández FE, et al. Chemical Variation and Biological Properties of the Essential Oil and Main Volatiles of Pimenta dioica Harvested in the Northern Highlands of Puebla, Mexico. *Chem Biodivers*. 2024 Nov 27:e202402843.

28 Moss M, Oliver L. Plasma 1,8-cineole correlates with cognitive performance following exposure to rosemary essential oil aroma. *Ther Adv Psychopharmacol*. 2012 Jun;2(3):103–113.

29 Priestley CM, Williamson EM, Wafford KA, et al. Thymol, a constituent of thyme essential oil, is a positive allosteric modulator of human GABAA receptors and a homo-oligomeric GABA receptor from Drosophila melanogaster. *Br J Pharmacol*. 2003 Nov 17;140(8):1363–1372.

30 Feng T, Hu Z, Song S, et al. The antioxidant and tyrosinase inhibition properties of essential oil from the peel of Chinese Torreya grandis Fort. *RSC Adv*. 2019;9:42360-42366.

31 Choi WS, Kwon MH, Kim YC. Inhibition effects of frankincense oil on skin aging (I): Focused on gross examination. *J Environ Toxicol*. 2008;23(20:119-27.

32 Padmakumari KP, Sasidharan I, Sreekumar MM. Composition and antioxidant activity of essential oil of pimento (Pimenta dioica (L) Merr.) from Jamaica. *Nat Prod Res*. 2011 Jan;25(2):152-60.

33 Singh P, Hayaramaiah RH, Agawane SB, et al. Potential Dual Role of Eugenol in Inhibiting Advanced Glycation End Products in Diabetes: Proteomic and Mechanistic Insights. *Sci Reports*. 2016 Jan 7;6:18798(2016).

34 Tabari MA, Tehrani MAB. Evidence for the Involvement of the GABAergic, but Not Serotonergic Transmission in the Anxiolytic-like Effect of Bisabolol in the Mouse Elevated plus Maze. *Naunyn Schmiedebergs Arch Pharmacol*. 2017;390:1041–1046.

35 Rombola L, Scuteri D, Watanabe C, et al. Role of 5-HT1A Receptor in the Anxiolytic-Relaxant Effects of Bergamot Essential Oil in Rodent. *Int J Mol Sci*. 2020 Apr 9;21(7):2597.

36 Weston-Green K, Clunas H, Naranjo CJ. A Review of the Potential Use of Pinene and Linalool as Terpene-Based Medicines for Brain Health: Discovering Novel Therapeutics in the Flavours and Fragrances of Cannabis. *Front Psychiatry*. 2021 Aug 26;12:583211.

37 Xu L, Han Y, Chen X, et al. Molecular mechanisms underlying menthol binding and activation of TRPM8 ion channel. *Nat Commun*. 2020;11(3790):(2020).

38 Moss M, Hewitt S, Moss L, et al. Modulation of cognitive performance and mood by aromas of peppermint and ylang-ylang. *Int J Neurosci*. 2008 Jan;118(1):59-77.

39 Komori T. Effects of lemon and valerian inhalation on autonomic nerve activity in depressed and healthy subjects. *Int J Essent Oil Ther*. 2009;3(1):3-8.

40 Chen M, Chang YY1, Huang S, et al. Aromatic-turmerone Attenuates LPS-Induced Neuroinflammation and Consequent Memory Impairment by Targeting TLR4-Dependent Signaling Pathway. *Mol Nutr Food Res*. 2018 Jan;62(2).

41 Hucklenbroich J, Klein R, Neumaier B, et al. Aromatic-turmerone induces neural stem cell proliferation in vitro and in vivo. *Stem Cell Res Ther*. 2014 Sep;5:100.

[42] Urasaki Y, Beaumont C, Talnot JN, et al. Akt3 Regulates the Tissue-Specific Response to Copaiba Essential Oil. *Int J Mol Sci*. 2020;21(8):2851.

[43] Akinyemi AJ, Adeniyi PA. Effect of Essential Oils from Ginger (Zingiber officinale) and Turmeric (Curcuma longa) Rhizomes on Some Inflammatory Biomarkers in Cadmium Induced Neurotoxicity in Rats. *J Toxicol*. 2018 Oct 8;2018:4109491.

[44] Horvath G, Horvath A, Reichert G, et al. Three chemotypes of thyme (Thymus vulgaris L.) essential oil and their main compounds affect differently the IL-6 and TNFα cytokine secretions of BV-2 microglia by modulating the NF-κB and C/EBPβ signalling pathways. *BMC Complement Med Ther*. 2021 May 22;21(1):148.

[45] Bachiega TF, de Sousa JP, Bastos JK, et al. Clove and eugenol in noncytotoxic concentrations exert immunomodulatory/anti-inflammatory action on cytokine production by murine macrophages. *J Pharm Pharmacol*. 2012 Apr;64(4):610-16.

[46] Carrasco FR, Schmidt G, Romero AL, et al. Immunomodulatory activity of Zingiber officinale Roscoe, Salvia officinalis L. and Syzygium aromaticum L. essential oils: evidence for humor and cell-mediated responses. *J Pharm Pharmacol*. 2009 Jul;61(7):961-67.

[47] Halder S, Mehta AK, Mediratta PK, et al. Essential oil of clove (Eugenia caryophyllata) augments the humoral immune response but decreases cell mediated immunity. *Phytother Res*. 2011 Aug;25(8):1254-56.

[48] Aldahlawi AM, Alzahrani AT, Elshal MF. Evaluation of immunomodulatory effects of Boswellia sacra essential oil on T-cells and dendritic cells. *BMC Complement Med Ther*. 2020 Nov 19;20(1):352.

[49] Young Living Essential Oils. Patent: Composition containing an essential oil product and method for using such to maintain normal levels of testosterone. Available at: http://www.google.com/patents/WO2014078590A1?cl=en.

[50] Almujaydil MS, Algheshairy RM, Alhomaid RM, et al. Nigella sativa-Floral Honey and Multi-Floral Honey versus Nigella sativa Oil against Testicular Degeneration Rat Model: The Possible Protective Mechanisms. *Nutrients*. 2023 Mar 30;15(7):1693.

[51] De Rubis G, Paudel KR, Manandhar B, et al. Agarwood Oil Nanoemulsion Attenuates Cigarette Smoke-Induced Inflammation and Oxidative Stress Markers in BCi-NS1.1 Airway Epithelial Cells. *Nutrients*. 2023 Feb;15(4):1019.

[52] Li YL, Du ZY, Li PH, et al. Aromatic-turmerone ameliorates imiquimod-induced psoriasis-like inflammation of BALB/c mice. *Int Immunopharmacol*. 2018 Sep 19;64:319-325.

[53] Ercin E, Kecel-Gunduz S, Gok B, et al. Laurus nobilis L. Essential Oil-Loaded PLGA as a Nanoformulation Candidate for Cancer Treatment. *Molecules*. 2022 Mar 15;27(6):1899.

[54] Hu Y, He Z, Zhang J, et al. Effect of Piper nigrum essential oil in dextran sulfate sodium (DSS)-induced colitis and its potential mechanisms. *Phytomedicine*. 2023 Oct;119:155024.

[55] Urasaki Y, Beaumont C, Talnot JN, et al. Akt3 Regulates the Tissue-Specific Response to Copaiba Essential Oil. *Int J Mol Sci*. 2020;21(8):2851.

[56] Chen G, Lv C, Nie Q, et al. Essential Oil of Matricaria chamomilla Alleviate Psoriatic-Like Skin Inflammation by Inhibiting PI3K/Akt/mTOR and p38MAPK Signaling Pathway. *Clin Cosmet Investig Dermatol*. 2024 Jan 8;17:59-77.

[57] Wei M, Liu F, Raka RN, et al. In vitro and in silico analysis of 'Taikong blue' lavender essential oil in LPS-induced HaCaT cells and RAW264.7 murine macrophages. *BMC Complement Med Ther*. 2022 Dec 6;22(1):324.

[58] Yang C, Zhang W, Xiang S, et al. Navel orange peel essential oil inhibits the growth and progression of triple negative breast cancer. *BMC Complement Med Ther*. 2024 Jun 14;24(1):233.

[59] Lee SY, Chen PY, Lin JC, et al. Melaleuca alternifolia Induces Heme Oxygenase-1 Expression in Murine RAW264.7 Cells through Activation of the Nrf2-ARE Pathway. *Am J Chin Med*. 2017;45(8):1631-1648.

[60] Alotaibi NM, Alotaibi MO, Alshammari N, et al. Network Pharmacology Combined with Molecular Docking, Molecular Dynamics, and In Vitro Experimental Validation Reveals the Therapeutic Potential of Thymus vulgaris L. Essential Oil (Thyme Oil) against Human Breast Cancer. *ACS Omega*. 2023 Dec 11;8(50):48344-48359.

[61] Mohanty D, Padhee S, Priyadarshini A, et al. Elucidating the anti-cancer potential of Cinnamomum tamala essential oil against non-small cell lung cancer: A multifaceted approach involving GC-MS profiling, network pharmacology, and molecular dynamics simulations. *Heliyon*. 2024 Mar 16;10(6):e28026.

[62] Tubachi SS, Rasal VP, Ugare SR, et al. Evaluation of Ylang Ylang essential oil on alcohol induced hepatotoxicity in rats. *Adv Trad Med*. 2022 Mar 4;191:1-14.

[63] Sepici A, Gurbuz, Cevik C, et al. Hypoglycaemic effects of myrtle oil in normal and alloxan-diabetic rabbits. *J Ethnopharmacol*. 2004 Aug;93(2-3):311-18.

[64] Kang P, Han SH, Moon HK, et al. Citrus bergamia Risso elevates intracellular CA (2+) in human vascular endothelial cells due to release of Ca (2+) from primary intracellular stores. *Evid Based Complement Altern Med*. 2013;2013:759615.

[65] Cherkaoui-Tangi K, Israili ZH, Lyoussi B, et al. Vasorelaxant effect of essential oil isolated from Nigella sativa L. seeds in rat aorta: Proposed mechanism. *Pak J Pharm Sci*. 2016 Jan;29(1):1-8.

[66] de Menezes IA, Moreira IJ, de Paula JW, et al. Cardiovascular effects induced by Cymbopogon winterianus essential oil in rats: involvement of calcium channels and vagal pathway. *J Pharm Pharmacol*. 2010 Feb;62(2):215-21.

[67] Khazaeli P, Chamani G, Mehrabani M, et al. Formulation and clinical evaluation of Myrtus Mucoadhesive paste in the treatment of recurrent aphthous stomatitis. *J Dental School Shahid Beheshti Univ Med Sci*. 2005;23:429-37.

[68] Coelho LS, Correa-Netto NF, Masukawa MY, et al. Inhaled Lavandula angustifolia essential oil inhibits consolidation of contextual- but not tone-fear conditioning in rats. *J Ethnopharmacol*. 2018 Apr 6;215:34-41.

[69] Adesina AB, Josephine OO. Inhibitory effects of the volatile oils of Callistemon citrinus (Curtis) Skeels and Eucalyptus citriodora Hook (Myrtaceae) on the acetylcholine induced contraction of isolated rat ileum. *Pak J Phark Sci*. 2012 Apr;25(2):435–39.

[70] Kang P, Ryu KH1, Lee JM, et al. Endothelium- and smooth muscle-dependent vasodilator effects of Citrus aurantium L. var. amara: Focus on Ca(2+) modulation. *Biomed Pharmacother*. 2016 Aug;82:467-71.

[71] Villareal MO, Ikeya A, Sasaki K, et al. Anti-stress and neuronal cell differentiation induction effects of Rosmarinus officinalis L. essential oil. *BMC Complement Altern Med*. 2017 Dec 22;17(1):549.

[72] Abd El-Aziz NK, Ammar AM, El-Naenaeey EYM, et al. Antimicrobial and antibiofilm potentials of cinnamon oil and silver nanoparticles against Streptococcus agalactiae isolated from bovine mastitis: new avenues for countering resistance. *BMC Vet Res*. 2021 Mar 31;17(1):136.

[73] Sivakumar L, Chellappan DR, Sriramavaratharajan V, et al. Root essential oil of Chrysopogon zizanioides relaxes rat isolated thoracic aorta - an ex vivo approach. *Z Naturforsch C J Biosci*. 2020 Oct 14;76(3-4):161-168.

[74] Shinohara K, Doi H, Kumagai C, et al. Effects of essential oil exposure on salivary estrogen concentration in perimenopausal women. *Neuro Endocrinol Lett*. 2017 Jan;37(8):567-572.

[75] Li H, Wang D, Chen Y, et al. β-Caryophyllene inhibits high glucose-induced oxidative stress, inflammation and extracellular matrix accumulation in mesangial cells. *Int Immunopharmacol*. 2020 Jul;84:106556.

[76] Pant A, Mishra V, Saikia SK, et al. Beta-caryophyllene modulates expression of stress response genes and mediates longevity in Caenorhabditis elegans. *Exp Gerontol*. 2014 Sep:57:81-95.

[77] Shrama C, Kaabi JMA, Nurulain SM, et al. Polypharmacological Properties and Therapeutic Potential of β-Caryophyllene: A Dietary Phytocannabinoid of Pharmaceutical Promise. *Curr Pharm Des*. 2016;22(21):3237-64.

[78] Shi X, Zhang W, Bao X, et al. Eugenol modulates the NOD1-NF-κB signaling pathway via targeting NF-κB protein in triple-negative breast cancer cells. *Front Endocrinol (Lausanne)*. 2023 Feb 27:14:1136067.

[79] Randhawa PK, Rajakumar A, de Lima IBF, et al. Eugenol attenuates ischemia-mediated oxidative stress in cardiomyocytes via acetylation of histone at H3K27. *Free Radic Biol Med*. 2022 Dec 13;194:326–336.

[80] Leong W, Huang G, Liao W, et al. Traditional Patchouli essential oil modulates the host's immune responses and gut microbiota and exhibits potent anti-cancer effects in ApcMin /+ mice. *Pharmacol Res*. 2022 Feb;176:106082.

[81] Nikolic B, Mitic-Culafic D, Vukovic-Gacic B, et al. Modulation of genotoxicity and DNA repair by plant monoterpenes camphor, eucalyptol and thujone in Escherichia coli and mammalian cells.*Food Chem Toxicol*. 2011 Sep;49(9):2035-45.

[82] Zhang G, Zhou X, Feng Q, et al. Nerolidol reduces depression-like behavior in mice and suppresses microglia activation by down-regulating DNA methyltransferase 1. *Neuroreport*. 2024 May 8;35(7):457-465.

[83] Jiang Y, He P, Sheng K, et al. The protective roles of eugenol on type 1 diabetes mellitus through NRF2-mediated oxidative stress pathway. *Elife*. 2025 Jan 10:13:RP96600.

[84] Kathem SH, Nasrawi YS, Mutlag SH, et al. Limonene Exerts Anti-Inflammatory Effect on LPS-Induced Jejunal Injury in Mice by Inhibiting NF-κB/AP-1 Pathway. *Biomolecules*. 2024 Mar 12;14(3):334.

[85] Rezyat SM, Dehpour AR, Motamed SM, et al. Foeniculum vulgare essential oil ameliorates acetic acid-induced colitis in rats through the inhibition of NF-kB pathway. *Inflammopharmacology*. 2018 Jun;26(3):851-859.

[86] Pandur E, Balatinacz A, Micalizzi G, et al. Anti-inflammatory effect of lavender (Lavandula angustifolia Mill.) essential oil prepared during different plant phenophases on THP-1 macrophages. *BMC Complement Med Ther*. 2021 Nov 24;21(1):287.

[87] Urasaki Y, Beaumont C, Workman M, et al. Fast-Acting and Receptor-Mediated Regulation of Neuronal Signaling Pathways by Copaiba Essential Oil. *Int J Mol Sci*. 2020 Mar 25;21(7).

[88] Yang H, Chen YX, Linghu KG, et al. 1,8-Cineole alleviates Nrf2-mediated redox imbalance and mitochondrial dysfunction in diabetes mellitus by targeting Sirt1. *Phytomedicine*. 2024 Oct 1;135:156099.

[89] doTERRA International. doTERRA Serenity Restful Complex Softgels. Available at: https://www.doterra.com/SG/en_SG/blog/scientific-research-lavender-peace-sleep-system, accessed March 3, 2025.

[90] doTERRA International. Frankincense Supplements and Inflammation. Available at: https://ichgcp.net/clinical-trials-registry/NCT06488417, accessed March 3, 2025.

[91] Falk-Flipsson A, Hagberg N, Lof AE, et al. Uptake, distribution and elimination of ??-pinene in man after exposure by inhalation. *Scandanavian J Work Environ Health*. 1990 Nov;16(5):372-8.

[92] Falk-Flipsson A, Lof A, Hagberg M, et al. d-limonene exposure to humans by inhalation: uptake, distribution, elimination, and effects on the pulmonary function. *J Toxicol Environ Health*. 1993 Jan;38(1):77-88.

[93] Falk A, Lof A, Hagberg M, et al. Human exposure to 3-carene by inhalation: toxicokinetics, effects on pulmonary function and occurrence of irritative and CNS symptoms. *Toxicol Appl Pharmacol*. 1991 Sep 1;110(2):198-205.

[94] Satou T, Takahashi M, Kasuya H, et al. Organ accumulation in mice after inhalation of single or mixed essential oil compounds. *Phytother Res*. 2013 Feb;27(2):306-11.

[95] Jager W, Buchbauer G, Jirovetz l, et al. Percutaneous absorption of lavender oil from massage oil. *J Soc Cosmet Chem*. 1992 Jan/Feb;43:49-54.

[96] Capetti F, Sgorbini B, Cagliero C, et al. Melaleuca alternifolia Essential Oil: Evaluation of Skin Permeation and Distribution from Topical Formulations with a Solvent-Free Analytical Method. *Planta Med*. 2020 Apr;86(6):442-450.

[97] Kohlert C, van Rensen I, Marz R, et al. Bioavailability and pharmacokinetics of natural volatile terpenes in animals and humans. *Planta Med*. 2000 Aug;66(6):495-505.

[98] Horky P, Skalickova S, Smerkova K, et al. Essential Oils as a Feed Additives: Pharmacokinetics and Potential Toxicity in Monogastric Animals. *Animals (Basel)*. 2019 Jun 13;9(6):352.

[99] Loureiro JP, da Rosa HS, de Araujo LS, et al. Andrographis paniculata Formulations: Impact on Diterpene Lactone Oral Bioavailability. *Eur J Drug Metab Pharmacokinet*. 2021 Nov 23;47(1):19–30.

[100] Horky P, Skalickova S, Smerkova K, et al. Essential Oils as a Feed Additives: Pharmacokinetics and Potential Toxicity in Monogastric Animals. *Animals (Basel)*. 2019 Jun 13;9(6):352.

[101] Vigushin DM, Poon GK, Boddy A, et al. Phase I and pharmacokinetic study of D-limonene in patients with advanced cancer. Cancer Research Campaign Phase I/II Clinical Trials Committee. *Cancer Chemother Pharmacol*. 1998;42(2):111-7.

[102] Chow HHS, Salazar D, Hakim IA. Pharmacokinetics of Perillic Acid in Humans after a Single Dose Administration of a Citrus Preparation Rich in d-Limonene Content. *Cancer Epidemiol Biomarkers Prev*. 2002 Nov 1;11(11):1472-76.

[103] Miller JA, Lang JE, Ley M, et al. Human breast tissue disposition and bioactivity of limonene in women with early-stage breast cancer. *Cancer Prev Res (Phila)*. 2013 Jun;6(6):577-84.

[104] Emami S, Azadmard-Damirchi S, Peighambardoust SH, et al. Liposomes as carrier vehicles for functional compounds in food sector. *J Exp Nanoscience*. 2016;9(11):737-59.

[105] Ricci C, Sparacino IM, Valerii MC, et al. Very-Low-Absorbable Geraniol for the Treatment of Irritable Bowel Syndrome: A "Real-World" Open-Label Study on 1585 Patients. *Nutrients*. 2025, 17(2), 328

[106] No author listed. Nanofiber in transmucosal drug delivery. Available at: https://www.sciencedirect.com/topics/pharmacology-toxicology-and-pharmaceutical-science/sublingual-route. Accessed March 10, 2025.

[107] Zhao Y, Wang C, Chow AHL, et al. Self-nanoemulsifying drug delivery system (SNEDDS) for oral delivery of Zedoary essential oil: Formulation and bioavailability studies. *Int J Pharmaceuticals*. 2010 Jan;383(1-2):170-77.

[108] Enin HA, Khalifa AS. Essential Oils Enriched Self-nano-emulsifying Systems for Effective Oral Delivery of Zaleplon for Improvement of Insomnia Treatment. *Indian J Pharmaceutical Ed Res*. 2022 Sep;56(4):1031-1043.

[109] European Medicines Agency. European Union herbal monograph on Carum carvi L., aetheroleum. 2015 Jul 7.

[110] German Commission E. Caraway oil (Carvi aetheroleum). 1990 Fen 1.

[111] European Medicines Agency. Community herbal monograph on Cinnamomum verum J.S. Presl (Cinnamomum zeylanicum Nees), aetheroleum. 2010 Jul 15.

[112] The complete German Commission E monographs, Therapeutic guide to herbal medicines, 1998.

[113] German Commission E. Eucalyptus oil (Eucalypti aetheroleum). 1990 Mar 13.

[114] Cosmetic Ingredient Review. Safety Assessment of Eucalyptus globulus (Eucalyptus) - Derived Ingredients as Used in Cosmetics. 2017 Nov 10.

[115] The complete German Commission E monographs, Therapeutic guide to herbal medicines, 1998.

[116] German Commission E. Niauli oil (Niauli aetheroleum). 1992 Aug 29.

[117] European Medicines Agency. OVERVIEW OF COMMENTS RECEIVED ON 'COMMUNITY HERBAL MONOGRAPH ON THYMUS VULGARIS L. AND THYMUS ZYGIS L., HERBA' EMEA/HMPC/234113/2006.

[118] Johnson SA. The Endocannabinoid System and Cannabis: The Perfect Partnership for Self-Regulation and Healing. March 7, 2019. Scott A. Johnson Professional Writing Services LLC., Orem Utah.

Index

bay laurel, 31, 40-42

BDNF, 110, 167, 183, 184, 215

bergamot, 9, 16, 23, 24, 25, 31, 43-47, 209

bile, 119, 120, 135, 178, 184, 185

bile acid metabolism, 119, 120

bile acid synthesis, 183, 184

biofilm, 13, 36, 38, 39, 54, 57, 63, 64, 70, 71, 78, 79, 81, 82, 99, 105, 106, 117, 120 ,121, 127, 128, 135, 144, 149, 153, 154, 156, 157, 169. 170, 177, 179, 186, 188, 189, 190, 191, 194

black cumin, 26, 31

black pepper, 23, 28, 31, 32, 47-50

black sage, 28

blood glucose, 31, 56, 57, 66, 67, 68, 69, 92, 126, 137, 173

blue tansy, 51-52

brainwave, 16, 86, 183, 197, 198

BRAF, 180 ,181

BRCA1, 8

butyrylcholinesterase (BChE), 14, 54, 124, 125, 131, 180

C

C-reactive protein (CRP), 20

cajeput, 25

calcium ion channels, 43, 44, 56, 84, 141, 158, 171

calcium signaling, 131

cananga, 23

cannabinoid, 23, 27-28, 32, 49,

caraway, 22, 213

cardamom, 52-55

cardiac troponin T, 136, 177

cardiovascular system, 142, 215, 216

carnitine palmitoyltransferase (CPT), 77

caspase, 44, 71, 96, 115, 116, 121, 127, 131, 134, 135, 167, 168, 181, 190, 191

cassia, 22, 56-58

catalase (CAT), 12, 37, 45, 49, 81, 84, 86, 96, 101, 138, 154, 155, 171, 176, 177, 192, 197

catnip, 24

CB1 receptor, 23, 27

CB2 receptor, 23, 27, 32, 49, 76, 198

CBR-1 receptor, 117

CD11b+, 147

CD4+, 20

CD45+, 147

CD8+, 20

cedarwood (Atlas), 22, 23, 58-60

cedarwood (Himalayan), 58-60

cedarwood (Virginia), 60-62

cell surface receptors, 30

cellular senescence, 193, 194, 195

cellular immunity, 73

central amygdaloid nucleus, 197

ceramide, 107, 108

choline, 170

choline esters, 131

cholinergic, 66, 125, 131, 139

cinnamon bark, 11, 12, 14, 24, 62-67, 213

cinnamon leaf, 62-67

circadian rhythm, 33, 195, 198

citron, 31

citronella, 31, 210

clary sage, 14, 29, 67-70

clove, 11, 12, 19, 32, 70-73, 210

collagen, 18, 25, 47, 49, 60, 61, 66, 69, 80, 89, 97, 101, 111, 112, 136, 151, 152, 159, 185, 187, 194

collagen I, 89, 111, 112, 151, 152

F

fatty acid synthase (FAS), 77, 119, 196

fatty acids, 38, 83, 138, 160

fennel (sweet/bitter), 24, 81-84, 213

fibrinogen, 20

fir (balsam), 85-86

fir (Siberian), 85-86

FOS protein, 37, 38

Frankincense, 21, 23, 24, 27, 33, 87-90

FRAP, 12, 42

furanocoumarins, 115, 209

G

GABA, 9, 15, 16, 23, 46, 80, 100, 110, 111, 112, 116, 117, 118, 120, 122, 128, 131, 140, 141, 143, 159, 162, 166, 168, 192, 193

GABA(A) receptors, 15, 117, 141, 159, 166

GABA A-BZD receptor, 141, 162

Gap43 gene expression, 170-171

gastric mucus, 52, 72, 117, 129

gene expression, 31-33, 66, 71, 99, 107, 108, 131, 135, 149, 151, 152, 169, 170 185, 195

geranium, 13, 15, 22, 32, 90-93

German chamomile, 16, 22, 31, 93-98

ginger, 10, 17, 22, 23, 24, 26, 98-101

glucokinase, 132, 180

gluconeogenesis, 31, 132, 173

glucose, 16, 17, 18, 30, 31, 33, 42, 56, 57, 63, 65, 66, 67, 68, 69, 92, 109, 115, 117, 118, 119, 123, 124, 125, 126, 132, 137, 138, 155, 172, 173, 180, 195, 196, 200

glucose-6-phosphatase, 132, 180

glucosidase, 15, 42, 65, 68, 69, 104, 138. 195, 196

glucosyltransferase, 71, 119

glutamate, 9, 45, 110, 120, 122, 162

glutamate receptors, 9, 45, 110

glutathione (GSH), 45, 49, 85, 96, 97, 101, 118, 171, 172, 176, 177, 197

glutathione peroxidase (GPx), 12, 113, 138, 139, 154, 155, 192, 197

glutathione reductase (GSH-R), 49, 86

glutathione-S-transferase, 50

glycogen, 126, 138, 173

glycogenesis, 31, 138

glycogenolysis, 31, 138

glycolysis, 138

glymphatic system, 17

grapefruit, 15, 22, 101-104, 209

H

H+/K+ ATPase, 53

H1 histamine receptor, 187

heart rate, 17, 25, 48, 86, 110 ,118, 133, 177, 202, 206

helichrysum, 15, 23, 104-108

heme oxygenase-1 pathways, 135

hemp, 23

hepatic GLUT4 and SREBP-1c, 132

hinoki, 24

hippocampus, 18, 38, 45, 62, 145, 163, 204, 206

histamine, 18, 39, 59, 95, 122, 186, 187, 188

histone deacetylases (HDACs), 32

HMG-CoA reductase, 113

hormone, 18, 19, 29-32, 46, 57, 68, 76, 82, 91, 93, 102, 104, 107, 109, 112, 114, 118, 123, 140, 142, 143, 161, 163, 164, 168, 178, 179, 184, 195, 204, 206, 214, 216, 217

hTRPM8, 79, 80

[229]

lymphocyte subset analysis, 20

lymphocyte, 20, 26, 88, 107, 200

LXR, 118, 119, 184, 185, 195

M

malate dehydrogenase, 192

malondialdehyde (MDA), 96, 171, 172

mandarin, 181-185

MAPK/ERK pathway, 31

marjoram (sweet), 22, 126-129

mast cells, 18, 58, 59, 94, 95, 112, 113

melanogenesis, 101

melatonin, 111

melissa, 22, 23, 26, 28, 130-133

metabotropic glutamate receptors, 9

mevalonate pathway, 113

microbiome, 14, 21, 48, 50, 81, 98, 100, 103, 112, 146, 154, 160, 182, 184, 185, 214, 216

mitochondria, 33, 43, 44, 47, 49, 95, 121, 130, 131, 149, 179, 190, 191, 192

MKP-1 phosphatase, 79, 80

MMP-1, 101, 151, 152

monocyte chemoattractant protein-1 (MVP-1), 152,

monokine induced by gamma interferon (MIG), 152

monoterpenes, 9-10, 32, 51, 167, 206, 208, 210, 211, 212

mRNA, 100,132, 163

mRNA expression, 100, 132

mucin, 24-25

MUC5AC, 25

MUC5b, 113

muscarinic pathway/receptors, 25-26, 131

muscle cells, 30, 45, 55, 126, 129, 139, 142, 158, 166, 173, 192, 193

muscular system, 214-215

myocardial infarction, 87, 90

myocardium, 39, 114

myeloperoxidase, 190

myrrh, 134-136, 175

myrtle, 31, 136-140

N

NADH-dehydrogenases, 192

natural killer cells, 20

neroli, 23, 31, 33, 129, 140-143

nerve growth factor, 162, 166

nervous system, 9, 16-19, 22, 25, 27, 37, 46, 48, 55, 76, 81, 82, 101, 104, 110, 112, 115, 116, 118, 122, 123, 124, 129, 133, 160, 162, 164, 166, 168, 174, 176, 193, 197, 200, 202, 206, 215, 217

neuraminidase, 66

neurite outgrowth, 163

neuroendocrine, 67, 147, 217

neurogenesis,161, 163, 184

neuroinflammation, 11, 17, 19, 45, 74, 76, 93, 100, 122, 195, 204

neuropeptide, 28, 111, 164, 188

neurotransmitter, 9, 16, 17, 18, 19, 23, 25, 27, 28, 37, 44, 46, 47, 48, 59, 60, 61, 62, 67, 68, 69, 72, 75, 77, 83, 84, 86, 89, 90, 97, 100, 107, 108, 110, 111, 112, 116, 118, 122, 128, 131, 143, 145, 155, 156, 158, 159, 162, 165, 166, 167, 170, 178, 180, 184, 185, 202, 203, 204

neutrophil, 11, 20, 26, 92, 97, 125, 139, 140, 188, 190, 200, 203

NF-κB, 10, 11, 19, 24, 32, 33, 36, 37, 45, 49, 52, 53, 54, 56, 57, 59, 63, 65, 68, 73, 80, 83, 88, 90, 99, 103, 109, 125, 128, 134, 135, 167, 168, 176, 177, 187, 194

niaouli, 32, 213

nicotine, 47, 48, 49

www.ingramcontent.com/pod-product-compliance
Lightning Source LLC
Chambersburg PA
CBHW061757260326
41914CB00006B/1145